BRUMBACK LIBRARY

3 3045 03192 8779

W9-ALL-989

$28.95
B/RODGERS
Rodgers, June Scobee
Silver linings

2/18

BRUMBACK LIBRARY
215 WEST MAIN STREET
VAN WERT, OHIO 45891

14
DAY
BOOK

PRAISE FOR *SILVER LININGS: MY LIFE BEFORE AND AFTER CHALLENGER 7*

In *Silver Linings: My Life Before and After Challenger 7*, June Scobee Rodgers chronicles an appealingly candid retrospect of determination and perseverance over adversity. This is an inspiring story of conquering hardships, encountering tragedy, demonstrating tenacity and, in creating the Challenger Centers for Space Science education, building something genuinely useful for society.

—NEIL ARMSTRONG

With firsthand knowledge of what it takes to overcome adversity June Scobee Rodgers tells her real life story before and after the *Challenger* tragedy. It's all about patriotism, passion, honor, selfless courage and heroism. She's my dear friend and will be an inspiration to all who meet her in the pages of *Silver Linings: My Life Before and After Challenger 7*.

—GEORGE H. W. BUSH, 41ST PRESIDENT OF THE UNITED STATES

June is truly one of my heroes. She turned grief into a positive force and helped us heal the wounds much more than we were able to help her and the other *Challenger* families. At a very difficult time in the history of our country, she was the backbone for us all, and for that, we'll always be grateful. *Silver Linings: My Life Before and After Challenger 7* is a true story of an American patriot.

—BARBARA BUSH, FORMER FIRST LADY OF THE UNITED STATES

I am reading *Silver Linings: My Life Before and After Challenger 7* and choking back the tears. This amazing story of June Scobee Rodgers's life could be a movie. I know firsthand how wonderful June and Don Rodgers represent the American spirit. My life is richer for knowing them and for honoring the *Challenger* crew.

—LEE GREENWOOD, ENTERTAINER, WRITER, MUSICIAN, SINGER & COUNCIL MEMBER FOR THE NATIONAL ENDOWMENT FOR THE ARTS

After my husband's tragic death onboard *Columbia*. . . . June lovingly shared with me her survival skills through the first days following her loss and gave me a lasting example of perseverance and tenacity. . . . In *Silver Linings: My Life Before and After Challenger 7*, readers will meet this remarkable woman. They will absorb her words of encouragement and faith and find courage to face challenges and tragedies on life's journey. Through this book she will become your friend as she has become mine. I promise.

—EVELYN HUSBAND THOMPSON, AUTHOR OF
*HIGH CALLING: THE COURAGEOUS LIFE AND FAITH OF
SPACE SHUTTLE COLUMBIA COMMANDER RICK HUSBAND*

Only someone as extraordinary as June Scobee Rodgers could transform the greatest tragedy in her life into a priceless gift of inspiration for millions of school kids around the world. *Silver Linings: My Life Before and After Challenger 7* carries an important lesson for us all: working in the service of others is great medicine. We're so lucky to have people like June in this world.

—SCOTT PARAZYNSKI, ASTRONAUT ON STS 66, 86, 95, 100, 120
CHAIRMAN, CHALLENGER CENTER

Silver Linings: My Life Before and After Challenger 7 is a brave and compelling book by a brave and compelling person. This is our story, America's story, and a story of the human spirit.

—KEVIN J. ANDERSON, *NEW YORK TIMES* BEST-SELLING AUTHOR

June Scobee Rodgers's moving story *Silver Linings: My Life Before and After Challenger 7* shows that God can take an ordinary life and use it in extraordinary ways. This book will give you hope, inspire you to dream big and challenge you to overcome whatever adversity you face. It is a must-read for those who feel they are limited by their circumstances.

—MARGARET PEALE EVERETT (DAUGHTER OF
NORMAN VINCENT PEALE)
CHAIRMAN, GUIDEPOSTS FOUNDATION

As I was reading *Silver Linings: My Life Before and After Challenger 7*, I found myself singing an old song by Jerome Kern, "Look for the silver lining when e'er a cloud appears in the sky. Remember somewhere the sun is shining, and so the right thing to do is make it shine for you." That's what June has been doing all of her life and especially since the *Challenger* tragedy in

1986. You will learn how she turned personal tragedy into triumph and literally turned some of her worst days into some of her best days. I enthusiastically recommend this book.

—**RON GLOSSER**, CEO BANK AND HERSHEY FOUNDATION

Silver Linings: My Life Before and After Challenger 7 is a powerful, moving, and enlightening book, by an amazing woman who had the strength and conviction to overcome a soul-crushing tragedy, to see through the darkness to find a silver lining that has helped over 8 million kids. Her story, this story, is an inspirational read that will change people's lives.

—**HARRY KLOOR**, WRITER, DIRECTOR, PRODUCER OF AWARD-WINNING FILM *QUANTUM QUEST*, PRESIDENT JUPITER 9 PRODUCTIONS

In this lovingly told story, June Scobee Rodgers takes readers on an inspirational journey through the harsh conditions of her childhood to reveal why she never gave up when faced with adversity. It tells how, as an adult, she was able to call on those childhood experiences to help a nation triumph over a great tragedy. *Silver Linings: My Life Before and After Challenger 7* draws together the rich history of the space program from the early days of von Braun rockets and the discoveries of NASA astronauts to the explorers of today and the emerging role of private space travel.

—**PETER H. DIAMANDIS**, CHAIRMAN/CEO, X PRIZE FOUNDATION CO-FOUNDER/VICE-CHAIRMAN, SPACE ADVENTURES, LTD.

PRAISE FOR *SILVER LININGS: TRIUMPH OF THE CHALLENGER 7*

The *Challenger* Seven lived in vibrant pursuit of a dream; and as long as we continue to pursue that dream, as long as we help it to touch the lives of our young people, then it can be said that we never truly lost those seven brave souls. In her book, *Silver Linings: Triumph of the Challenger 7*, and by her everyday example, June Scobee Rodgers proves that this dream does indeed endure: the adventure of exploration and discovery embodied in the *Challenger* crew awaits those who have the will to make it happen.

—**GEORGE H. W. BUSH**, 41ST PRESIDENT OF THE UNITED STATES

June Scobee Rodgers is a remarkable woman, and her book tells an inspiring tale of triumph arising from tragedy. America's space program has always been about discovery—discovery of our universe, our world, and most importantly, ourselves. Although the flight of the *Challenger* ended tragically, June Scobee Rodgers never gave up. June has continued the *Challenger*'s mission of discovery and education with courage, faith, and a tireless energy. Her moving story will touch all readers.

—JOHN GLENN, FORMER ASTRONAUT AND SENATOR

Silver Linings: Triumph of the Challenger 7 is the true and inspiring story of the triumph of faith in the face of overwhelming defeat. . . . From lowly beginnings and through wrenching struggle, June Scobee Rodgers has proven herself to be one of God's great winners. Her story has touched my heart and renewed my conviction that God can turn all scars into stars. May her story here lead you beyond all your tribulations into triumph.

—ROBERT SCHULLER

PRAISE FOR THE CHALLENGER CENTER FOR SPACE SCIENCE EDUCATION AND CHALLENGER LEARNING CENTERS

Inspiring. Exploring. Learning. It's our mission. That's hard to beat!

—JOHN GLENN, FORMER ASTRONAUT AND SENATOR

The mission of the Challenger Center is to spark in our young people an interest—and a joy—in science. A spark that can change their lives—and help make American enterprise the envy of the world.

—GEORGE H. W. BUSH, 41ST PRESIDENT OF THE UNITED STATES

What the Challenger Center has done with respect to educating America's youth is truly commendable. I salute you.

—GENERAL COLIN POWELL

Praise for the *Star Challengers* Series

The *Star Challengers* series takes young readers up into space, onto the moon, and to the boundaries of their imaginations. It's the next best thing to being there.

—NEIL ARMSTRONG

Space exploration is a great adventure that benefits all mankind. The *Star Challengers* books inspire young readers with that sense of adventure, introducing them to a new universe of exciting possibilities.

—BUZZ ALDRIN, ASTRONAUT, APOLLO 11

The *Star Challengers* adventure stories could help to inspire a whole new generation of young women to value science and seek careers in high-tech, engineering and space exploration. These teenage Star Challengers team up in their quest to find innovative solutions to help them solve problems using real out-of-this-world science.

—SALLY RIDE, ASTRONAUT

The *Star Challengers* with their Commander Zota "boldly go into the future" to bring great science fiction adventures to their readers. What a wonderful way to expand young imaginations.

—LEONARD NIMOY, ACTOR AND DIRECTOR

In no other regime do reality and fiction seem to meet as commonly as in space. No wonder young (and old) people are inspired and excited when reading the *Star Challenger* series. It would be a great item to take along on one's next interplanetary voyage.

—NORM AUGUSTINE, RETIRED CHAIRMAN AND CEO, LOCKHEED MARTIN CORPORATION

There's a reason why the best science fiction takes place in space. It's the only true frontier left. Kids know this. So too does the *Star Challengers* series. Therein is the magical recipe to ensure a future in space for the rest of us.

—NEIL DE GRASSE TYSON, ASTROPHYSICIST, AMERICAN MUSEUM OF NATURAL HISTORY

The Challenger Centers continue to be a champion for the future. Young readers will readily identify with the *Star Challengers* characters. The future needs them, and they will respond—in wonderful ways.

—**BARBARA MORGAN**, NASA'S FIRST EDUCATOR ASTRONAUT

Ad astra! To the stars! By the way of good stories! Thank you for *Star Challengers*, Rebecca Moesta and Kevin J. Anderson.

—**CLAY MORGAN**, AUTHOR OF *THE BOY WHO SPOKE DOG*

June Scobee Rodgers is a woman on a mission and that mission continues to expand. By nature June is an encourager and an inspiration. She has worked to bring renewed interest in science education and space travel through the *Star Challenger* series which will help feed young, curious minds with the possibilities that await them in the future. I can't recommend these books more. Move over *Avatar* . . . here comes Commander Zota.

—**DEBBIE MACOMBER**, #1 *NEW YORK TIMES* BEST-SELLING AUTHOR

Space may be the final frontier according to *Star Trek*, but if our message to the next generation is to reach for the stars, then the *Star Challengers* series is a great place to start. Our future survival will depend upon how our young students meet the challenge of combining science, engineering, mathematics, and imagination.

—**LEE GREENWOOD**, ENTERTAINER, WRITER, MUSICIAN, SINGER & COUNCIL MEMBER FOR THE NATIONAL ENDOWMENT FOR THE ARTS

What if Earth's future rested on the shoulders of five ordinary teens living in present times? And what if a visitor from the future accompanies them through time and space for the adventure of their lives? I was charmed by the premise of *Star Challengers*, a new and innovative series geared to teen readers especially drawn to science and space technology. . . . I found the stories infused with nail-biting adventure, romance, and plausible science. Skip the vampires! Don't miss this thought-provoking series presented by June Scobee Rodgers and the Challenger Center for Space Science Education and written by award-winning, international best-selling authors Rebecca Moesta and Kevin J. Anderson.

—**LURLENE MCDANIEL**, BEST-SELLING AUTHOR

SILVER LININGS

MY LIFE BEFORE AND AFTER *CHALLENGER 7*

JUNE SCOBEE RODGERS

B
RODGERS

Smyth & Helwys Publishing, Inc.
6316 Peake Road
Macon, Georgia 31210-3960
1-800-747-3016
©2011 by June Scobee Rodgers
All rights reserved.
Printed in the United States of America.

The paper used in this publication meets the minimum requirements of
American National Standard for Information Sciences—
Permanence of Paper for Printed Library Materials.
ANSI Z39.48–1984. (alk. paper)

Library of Congress Cataloging-in-Publication Data

Rodgers, June Scobee.
Silver Linings: My Life Before and After *Challenger 7* / by June Scobee Rodgers.
p. cm.
Includes bibliographical references and index.
ISBN 978-1-57312-570-3 (hardcover : alk. paper)
1. Challenger Center for Space Science Education
2. Astronautics—Study and teaching (Elementary–Secondary);
Science, Technology, Engineering, Math
(STEM) Education—United States.
3. Rodgers, June Scobee.
4. Christian biography. I. Title.
TL847.C43R63 2010
629.4072'073—dc22

2010042751

Published in association with
Books & Such Literary Agency, 52 Mission Circle, Suite 122, PMB 170, Santa Rosa, CA 95409-5370
www.booksandsuch.biz

Disclaimer of Liability: With respect to statements of opinion or fact available in this work of nonfiction, Smyth & Helwys Publishing Inc. nor any of its
employees, makes any warranty, express or implied, or assumes any legal
liability or responsibility for the accuracy or completeness of any information
disclosed, or represents that its use would not infringe privately-owned rights.

High Flight

Oh! I have slipped the surly bonds of Earth
And danced the skies on laughter-silvered wings;
Sunward I've climbed, and joined the tumbling mirth
of sun-split clouds—and done a hundred things
You have not dreamed of—wheeled and soared and swung
High in the sunlit silence. Hov'ring there,
I've chased the shouting wind along, and flung
My eager craft through footless halls of air.
Up, up the long, delirious, burning blue
I've topped the wind-swept heights with easy grace
Where never lark nor even eagle flew—
And, while with silent lifting mind I've trod
The high untrespassed sanctity of space,
Put out my hand, and touched the face of God.

— John Gillespie Magee, Jr.

Dedication

To my family . . .

My brothers, who shared my childhood
and the memory of our parents

My children, Kathie and Rich, with whom I share
the joys and memories of their father, Dick Scobee

And to Don Rodgers, in gratitude for my new life
that includes nine wonderful grandchildren

To the *Challenger 7* families, for our shared loss and joy of friendship

To the Challenger Centers and teachers everywhere, who help launch
young people toward discovery and achievement

Acknowledgments

How does one thank a cast of thousands? My appreciation for contributions to this book and to the Challenger Center efforts reaches around the world. My gratitude for both individuals and organizations that came to the rescue of the *Challenger 7* families grows each day. For the people who made a significant impression on my young life and believed in me when all others thought I was a lost cause, I am forever indebted. For family and friends who still like me in spite of my flaws, I am ever thankful.

Specifically, I am grateful to the following people who made this book possible: Wendy Lawton at the Books and Such Agency, publisher Lex Horton, and everyone at Smyth & Helwys. They have been a joy to work with, especially my editor, Leslie Andres, who blessed this book with her creative talents, magic, and dedication. To the network of Challenger Centers' staff, flight directors, and educators, who made so many wonderful contributions to the *Silver Linings* story that we had to create this sequel. In particular, I want to thank our Challenger Center Board of Directors and the *Challenger 7* families, who have contributed mightily with leadership for the past twenty-five years, and to the teachers around the world who inspire their students to reach for the stars.

Individually, I want to thank my family for encouraging and inspiring this book. Their love has given me wings. At the Challenger Center, Pam Peterson, Cheri Winkler, Tariq Shazlee, and Rita Karl have been logistical magicians. Astronauts and chairmen Bill Readdy, Scott Parazynski, and Joe Allen have been champion leaders and supporters, as have our other chairmen and presidents, including Dan Barstow and our entire team from Challenger Center. My personal photo doc, Tracey Knauss, helped me revive several older photographs and create copies for the book. My daughter, Kathie Fulgham; Kristen Youmans; Harvey Yazijian; Gwen Griffin; Lurlene McDaniel; Jan Silvias; and Leslie Andres all helped to reassure me or help me find the right words when I was stumped. (I have included many conversations from memory and thus used creative license to recreate them when necessary.) I am also grateful to former astronaut Bob Crippen, who gave me invaluable help with space shuttle history, and to Keith Cowing, who has carried the Challenger Center to new heights and distances. I am immensely gratified for all the wonderful people who wrote heartfelt words of praise as endorsements for this book encouraging others to read its pages, especially my precious friend Debbie Macomber (with over 60 million books in print) for writing the foreword.

Most of all, I want to acknowledge the readers who pass this book on to others—to a teenager living a life of hopelessness, to an adult widowed too soon, and to anyone seeking a path to faith and the joy of a new found love of life and adventure.

And as always, we are indebted as a nation for the space pioneers who blazed a trail of exploration and discovery, and to the young people whom we hope to inspire to new adventures that can create opportunities for a better life upon this planet and beyond for all the generations to follow.

June Scobee Rodgers

CONTENTS

FOREWORD

Few people I've met in my life have impressed me more than June Scobee Rodgers. We came to know each other through our affiliation with Guideposts, started in 1945 by Norman Vincent Peale. (*Guideposts* magazine is one of North America's most popular periodicals).

It seems fitting that I would meet June through an organization whose goal is to encourage and inspire, since June Scobee Rodgers is one of the most positive, affirming women I have been privileged to know. Our friendship has blessed my life, and her story will bless you as well.

Most people recognize June as the widow of Dick Scobee and will instantly recall the tragedy of the 1986 *Challenger* explosion. The event riveted America and the world. Many people have compared the public outpouring of shock and grief to the emotions we experienced at the assassination of President John F. Kennedy.

In a single moment on that winter day in 1986, June's world forever changed. However, the fate of the *Challenger* is only part of her story.

Silver Linings: My Life Before and After Challenger 7 is about a girl who rose above her circumstances, who married young and raised her siblings while starting a family of her own. It's also the story of a man who dreamed of flying airplanes, never realizing that he would one day travel in space. As you become acquainted with June through her book, you will appreciate as I did that this is an extraordinary woman whose caring heart and godly wisdom will make a lasting impression on you.

Born into poverty and adversity, June faced many difficulties from an early age—and faced them head on. After the tragic loss of her husband, Dick, she carved out a future for herself and her children. But she didn't limit her efforts to her immediate family. She, along with the other spouses of the *Challenger* crew, also made an important contribution to the community of children in the United States and worldwide. They founded the Challenger Learning Centers as a positive and practical way of honoring the *Challenger* crew. With June as their leader, the *Challenger* families have completed the mission their loved ones set out to accomplish: educating and motivating schoolchildren, particularly in mathematics, technology, and the sciences.

The Challenger Learning Centers now number 50 in the United States and across the world. Because June and the other family members looked for the silver linings and golden opportunities in this devastating situation, more than 400,000 students and 40,000 teachers participate annually at the Challenger Learning Centers. Over eight million students have walked away inspired and encouraged, challenged (exactly the right word here!) to look beyond our world and into the frontier of space travel.

As you read June's story, you'll see that she is a woman of deep personal faith. Guided by love and the sure knowledge of God, June has taken the tragedies that have come into her life and refused to be defeated by them. Instead of drowning in a river of sorrow, she panned there for gold.

May your journey with June through the pages of this book open your eyes and your heart to the challenges in your own life, so that you, too, will find the silver linings and the golden opportunities.

Debbie Macomber
#1 *New York Times* Best-Selling Author

PREFACE

"All the world's a stage" is a phrase that begins a monologue from William Shakespeare's comedy *As You Like It*. The speech compares the world to a stage and life to a play and lists the seven stages of a man's life, sometimes referred to as the seven ages of man: infant, schoolboy, lover, soldier, justice, sage and on to aged or second childhood.

In this story, I attempt to focus on three stages of my life. The first begins with my starry-eyed and turbulent childhood and a nightly routine to wish upon the first star. Starved to learn more about the stars, rockets and space ships, and find an escape from my impoverished reality, I read science fiction and the local newspaper about the real world of astronomy and the scientists building rockets just up the highway from our rural Alabama farmhouse.

When I was nine years old, I stumbled upon a strategy to teach myself about positive thinking and practical techniques for overcoming life's problems. I discovered it all in a book written by Norman Vincent Peale, a gift to my mother along with some lovely tangerines. With an expanding awareness of the heavens, I learned that God who created the universe answers prayers. With newfound faith in His unwavering love, I discovered my own inner compass and began to turn my hopeless life in a new direction. I endeavored to accept problems as challenges in a childhood sprinkled with death, divorce, mental illness, poverty, and rejection.

In the second stage, my life was less turbulent and even idyllic at times. Dick Scobee and I married as teenagers and worked hard to support each other's academic and professional careers. Like many, though, we suffered personal and professional setbacks, self-doubt, failures, and disappointments, but our lives together were dotted by many successes and accomplishments especially with the shared joy of two wonderful children.

Twenty-five years into our marriage, we had both realized our dreams—Dick's dream to fly airplanes and mine to teach. Then came the exciting announcement that brought our dreams together: Dick would command the STS 51-L flight and Christa McAuliffe, the teacher from New Hampshire, would join the crew. We were overjoyed! Almost immediately, flight 51-L became known as the "Teacher in Space" mission.

Finally, launch day for the twenty-fifth shuttle flight arrived. It was January 28, 1986. At dawn that brisk morning at Kennedy Space Center, the *Challenger* crew, led by my husband, climbed aboard their Space Shuttle. It was so cold that icicles hung from the shuttle and gantry. Poised for lift-off, the orbiter sparkled in the cool brightness of the Florida sun. The engines ignited a thunderous roar, and the ground shook with the raw power of 6.5 million pounds of thrust. As the shuttle lifted off the launch pad, the air crackled with the roar of engines and cheers from the crowd.

We shared staunch faith and lofty dreams with the other *Challenger* families, and our spirits soared as we gathered to watch the launch—but then the unthinkable happened. Standing together at Kennedy Space Center, we watched the shuttle rip apart and the orbiter explode in the sky. Like our hearts, it shattered into a million pieces. We stood in numb disbelief; shock; silence.

The moments that unfolded before us on that cold January morning will forever be etched in our memories and that of our nation. Years later, my daughter, Kathie, wrote a letter to the children of the *Columbia* astronauts who died. She expressed our feelings at the time her father died aboard *Challenger*. She wrote, "My father died a hundred times a day on televisions all across the country. And since it happened so publicly, everyone in the country felt like it happened to them, too. They wanted to say goodbye to American heroes. Me? I just wanted to say goodbye to *my* Daddy."

With the "Teacher in Space" tragedy and loss of my beloved husband, a dark cloud settled over our families and our nation that had lost seven space pioneers. To help overcome our own loss and also that of a nation of children still waiting for lessons from space, I returned to my childhood strategy to overcome tragedy. Our families came together to create the Challenger Center non-profit organization to continue the *Challenger* mission for students. Before we knew it, a bright light radiated from behind the clouds of grief to reveal silver linings of hope and optimism.

In the third stage, my life has grown more complex, but it is also filled with the joy of many new opportunities for discovery and outreach to advance science education and to help others seeking to overcome adversity. Twenty-five years after the *Challenger* tragedy, the mission continues around the world in the creation of the Challenger Centers, now positioned as a national leader in science education for youth, introducing more than eight million students to a new universe of exciting possibilities.

Questioned by reporters over the years to explain what in my youth taught me how to turn the sorrow of *Challenger* into a legacy of hope, I was inspired to expand my original story about the *Challenger 7* and our tribute to them to include my personal story of faith. To help me answer their questions, I sometimes expressed my delight in the book that I read as a child about how to set in motion a strategy to overcome adversity. Guided by love and the sure knowledge of God, I learned that to fulfill dreams of a more positive future, I had to accept my problems as challenges to triumph over troubles.

This age also afforded me the luxury of time to write and publish the stories I had dreamed of telling about new wonders of our universe through science fiction, but this time as a co-author in the *Star Challengers* series. These novels that I developed with best-selling authors Kevin Anderson and Rebecca Moesta take young readers to a future Moonbase, a space station, and an asteroid probe—all simulations offered by the Challenger Centers. The project has received enthusiastic endorsements from Neil Armstrong, Buzz Aldrin, Sally Ride, Leonard Nimoy, Neil Tyson, and more.

Through my involvement on the Vatican Observatory Foundation Board, I learned firsthand how science and faith are compatible—the Church has not only supported astronomical research but has seen the study of the heavens as a way of getting to know the Creator. Also, recalling that the British author C. S. Lewis used Mars as the setting to write about faith in his novel *Out of the Silent Planet* motivated me to write stories about courageous kids and their "out of this world" adventures.

Silver Linings: My Life Before and After Challenger 7 is more than a book about a ragged teenager who struggles to overcome adversity, or a widow of the *Challenger* commander. It's a coming of age story; it's an intimate story about a woman in love and the power of that love to prevail over the *Challenger* tragedy. It's a story of the passion and faith that have guided my quest toward self-realization and have fueled my desire to make a long-lasting difference in the lives of others.

As I approach yet another age, I am ever grateful for God's divine intervention to a newfound love and life. After a chance meeting with Lt. Gen. Don Rodgers at an Easter sunrise service at Arlington National Cemetery and our subsequent marriage later that year, I discovered even greater happiness with the blessings of our family growing beyond my own dear children to the joy of nine delightful grandchildren.

Reflecting on my life, I recall my childhood depths of pain and rejection when kids would not sit next to me on the school bus, but also the joy of inclusion when kings and queens, presidents and other dignitaries invited me to dine with them at their tables. I've loved the embrace of steadfast friends who welcomed my company in spite of my flaws, and I've enjoyed the satisfaction of having accomplished a few goals and completed some projects. No reward, though, is as heart-thumpingly wonderful as being a teacher and making a difference in the lives of my students.

Can the sum of one's life be calculated? Or even measured in its worth? Life is filled with mountain top highs and valley lows, but truly one of the greatest joys I've ever witnessed is the mighty power of God's love, and the simple pleasure of being surrounded by a loving family who treasure and value time spent together—not necessarily the rip-roaring celebrations, but the less obvious moments of watching them through eyes that glisten with pride as they grow and learn and accomplish meaningful pursuits in their own lives and become loving parents themselves.

As I approach the last stages in my life, I find unexpected aspiration and opportunity; although I've collected outward signs of aging with wrinkles, my goal now is to nurture my feisty little girl enthusiasm and spirit to prevent wrinkles from ever forming on my soul.

June Scobee Rodgers
November 2010

PART 1

STORM CLOUDS

Chapter 1

EARLY MORNING CLOUDS

BANKRUPT AND MOVING ON

The summer I was thirteen, Daddy declared his small carpentry business was bankrupt. It was 1955 in Birmingham, Alabama, and my mother had just returned home from a long stay in the mental hospital.

"What does 'bankrupt' mean?" I asked. I learned it meant that Daddy had to clear out his office—and that we were broke.

Our family of six had been living in a two-bedroom garage apartment while Daddy built us a modest home on the same lot. But with the bankruptcy and no job, we lost the apartment, the unfinished house, and some of our furniture and possessions. Our parents decided we should leave the city and start a new life in the rural town of Odenville, just a pink dot on the map on the road to Gadsden, Alabama.

I remember crying in frustration when Daddy gave us the news that we would have to move so he could find another job. We had moved so many times and I had changed schools so often that I didn't know who I was or where I belonged.

In a huff, I ran barefoot from the house. Letting the front door slam behind me, I hurried down the driveway and out onto the hot asphalt toward my friend Jane's house. I had to tiptoe down the street for two long blocks, skip across the burning pavement, and then jump the curb to a hard, worn path of gravel that pricked my feet. When I arrived at Jane's front door, my blistered and bruised feet made me temporarily forget about the ache and anger in my head.

I knocked impatiently, and my eyes filled to overflowing just as Jane opened the door.

"What's wrong?" she asked. "Come in!"

I hobbled into her house and flopped down in an overstuffed chair.

"What happened to your feet? Why are you crying?" she pleaded, not waiting for answers.

I explained about my feet and then blurted out the devastating news. "Daddy's gone bankrupt, and we have to move out to the country in a few days."

Jane began crying too. "But we were going to go to high school together. You and me and Joe—the dynamic trio!"

We both cried together, realizing the unfairness of it all. Jane knew I'd moved more than twenty times since childhood and had been to at least fifteen different schools around Alabama, Texas, and Florida—sometimes with my whole family, and sometimes just with Mother, when she left Daddy and ran away with us. After all that, I had finally found a truly good friend, and the idea of moving *again* broke my heart.

"You'll be my best friend always," Jane promised. Then she wrinkled her freckled nose. "It must be awful to have to move so many times. We've lived in this same house all my life. My dad says he'll never move because we have roots here."

Roots? I couldn't imagine it. We'd never lived anywhere long enough to collect dust, much less fertile soil.

Jane and I talked, laughed, and wept for a while longer. Finally, I stood up to go.

"Please come visit us in Odenville!" I begged Jane.

"Of course. Soon we'll be driving cars, and we can visit each other all the time," she assured me.

As I stepped toward the front door, Jane reached into the bottom of her coat closet and handed me some old flip-flops, saying, "Here, you need these more than I do." After another hug and more tears, I left.

Back at the house, Daddy and my younger brothers, Lonnie and Johnny (ages ten and seven), were at the kitchen table scanning the want ad section of the newspaper. Daddy found a house for sale with land and a barn. "Hey, look at this one," he said. "It's a farm with a barn so we can have farm animals and grow a garden. We'll survive just fine, and we can learn how to live off the land."

Of course I had dozens of questions. "What's the house like? Will I get to have my own bedroom like in the house you were building? Will we have horses and cows? Where's the high school?" He hushed me, reminding me to be patient because we'd learn the answers soon enough. I wondered what it meant to "live off the land." Years earlier, when my mother had run off with us and left Daddy behind, we'd had to live like homeless people; I never wanted to live like that again—feeling vulnerable at roadside parks or fearful on city streets. Just the memory gave me cold shivers.

BIG DREAMS

Within days, we packed our battered pickup truck with a few household items and drove to a strange place for a fresh start once again. After we turned off the main highway in Odenville, we drove for two miles over a dirt road where a stream of passing cars stirred up a pink cloud of dust. As we went around a bend, Daddy turned off the rutted road and pulled up to a dilapidated wooden frame house with a hard-packed dirt yard. The house was small, dusty, and weather-beaten, and I felt sorry for the people who lived there. Daddy glumly announced, "We're home."

Except for a few undersized trees, a small barn with a chicken coop, and an outdoor toilet, our yard was an acre of the same barren red ground as the road. We had no neighbors for half a mile on either side, but I was excited to learn that acres of lush pine trees and winding trails surrounded our land. Even before we unloaded furniture and boxes from the pickup truck, Lonnie, Johnny, and I made the best of a bad situation, running through the shaded woods, our shoes crushing the pine needles and releasing a fresh woodsy scent.

Racing back to the house, skipping across broken boards on the covered back porch, I went into the kitchen and found Mother pacing the floor with our baby brother, Lee, who was just a year old. When she turned to face me, I saw her eyes were red from crying, and the baby was whimpering.

"What's wrong?" I wanted to know.

"This is the kitchen," she said, gesturing around her.

I saw only one small cupboard, a wood stove for cooking, and a bench. There was no sink or running water. Quickly, I took the baby from her arms and suggested we look at the rest of the house. While we talked about where to place the sofa and chair in the living room, Daddy set up the crib in the front bedroom so I could put the baby down for his nap.

"This must be your and Daddy's bedroom," I said. "Where's mine?"

We soon realized there was just one other bedroom. I'd have to share a room with my two brothers. There was no indoor bathroom. Instead, we placed a chamber pot under a bench on one side of the kitchen. On a shelf above the bench were a wash pan, a bar of soap, and a mirror where Daddy could shave and we could spot bathe. I learned to wake up first each morning so I could wash in private.

In the summer, we'd be able to go to the creek to wash in the spring that bubbled up fresh cool water. We could also collect a barrel of water from the clean side at the spring's mouth and haul it back in a wagon pulled by a mule.

Back at the house, we'd hoist the barrel onto the rickety back porch and shove it into its place in the corner. We dipped buckets of water out of the barrel and placed them on a large wooden shelf. A dipper hung on a nail in the support post for the roof. When the wind blew, the dipper cup would swing like a pendulum and clank against the wall. All the family drank from that dipper, but we set aside an old tin cup for people who visited.

We would wash dishes with that water, too, heating it in a tub on the wood stove and adding detergent. We filled another pan with cold water for rinsing. I remember the neighbor's skinny hound dogs waiting at the edge of the house, mouths drooling, for us to dump the dishwater and scraps of food. When we tossed it way out onto the ground, the dogs lapped up the sudsy water like it was a feast.

The day we moved into the house, Daddy and my brothers brought in our white enamel kitchen table, some chairs, and the refrigerator, which he plugged into the wall outlet. The motor gave a spurt and then began to whir.

"Look at this! We have electricity!" he declared. He screwed a light bulb into a socket hanging from a dingy electrical cord in the middle of the ceiling. We had camped out with lanterns before, so we knew to appreciate electricity.

Even so, I felt like crying along with my mother, whose tears began to flow freely again. I wanted to shout, *I'm no longer a little girl! I'm growing up! I need privacy!*

Instead, I asked, "Daddy, you can build us a bathroom, can't you? And make a cabinet with a sink?"

"Yes," I heard him struggle to say. Then he mumbled, "We'll dig a well after we drill for oil and strike it rich."

Mother turned to answer in a huff, so I took her arm and encouraged her to help me unpack the boxes and set up the beds before nightfall. Instead, she dug in her purse for a cigarette.

Daddy switched gears to a survival mode. He motioned Mother and me to sit on the rickety back stairs and pointed into the distance as he lit his own cigarette. I already knew he had big plans to make a garden, buy a cow for milk, and get chickens for eggs.

Looking out to the horizon, he asked me, "June Bug, can't you and your mother just see a sprawling big house over there on that hillside beyond where we'll plant the garden? You won't have to share a bedroom with your brothers, and we'll have running water and bathrooms and space heaters for every room."

Daddy was always a dreamer. Even though we often had to move two or three times a year to avoid bill collectors or because mother was ill and ran away with us, Daddy had big dreams. I figured that a house in any condition was better than some places where we had lived. Most of all, I was simply grateful when we lived together as a family, with both parents caring for us.

That night, it seemed as though I was at the end of the world in some kind of century-old time warp. Confused and bewildered, I smiled at Daddy for sharing his dreams, but I wondered how we'd gotten to these shabby backwoods of Alabama. How had we gone back in time, transported to an earlier era with no indoor plumbing and no heat?

I was tired of dreams that seemed unattainable, but I encouraged Daddy for Mother's sake.

"We'll help you build the house and plant the garden," I said, "but it's just so hard living like this." I thought about the article I had seen in the paper about Chuck Yeager. He had set a new speed record flying the Bell X-1A to Mach 2.4. I had read about X-15 pilots flying at altitudes high enough to earn "astronaut" ratings, and I wasn't too young to realize we had moved backward while the rest of the country moved forward.

Mother stubbed out her cigarette under her shoe and scoffed, "You are both dreamers with not a practical bone in your bodies. June has her head in the clouds, and you"—she looked at Daddy—"you're just a . . . ." Her voice trailed off as she turned to go back into the house.

We had no television, so we subscribed to a daily newspaper and listened to the radio to keep up with life beyond our rustic existence. DJs on the local radio station regularly played Elvis Presley songs. That summer, he had the number one single, "Don't Be Cruel," and on the flip side was "Hound Dog." *How appropriate,* I thought. In the distance, I heard the neighbor's hound dogs howling every night.

Daddy and I were the only ones who read the daily *Birmingham News* that a driver tossed from his car onto our front yard every afternoon. My brothers liked the comic page, but Daddy and I talked about baseball scores (the Yankees won the World Series) and politics (Dwight Eisenhower was reelected president). Sometimes I read aloud any major news stories.

One such evening after we ate dinner and cleared the dishes, I spread the newspaper across the kitchen table looking for news from the "outside" world. There on the front page, the space age came to life for me. I read from the article and asked Daddy, "Did you know that rocket research is being moved to just outside of Huntsville?" Huntsville was a cotton town in north Alabama, not far from Odenville. I continued reading. "It says von Braun, the rocket scientist, and his team have started working on the V-2 rocket called the Redstone designed to launch America's first satellite."

Daddy grinned at my excitement. "You're getting a bit spacey about those planes and rockets, aren't you?" He folded the paper and reminded me, "The sooner you heat up a bucket of water, the easier it will be to wash those dishes. Your mother is feeding the baby and doesn't feel well. Help your brothers wash up and get ready for bed, and then you can do your homework."

I knew he was tired from his farm chores and the hour drive to and from Birmingham for odd carpentry jobs, but I still grumbled at my chores. My brothers' only jobs were to haul water from the creek and pick up coal left over from the mines.

I fetched water from the barrel on the back porch, heated it on the wood stove, and washed the dishes, all the while daydreaming about the scientists and rockets in Huntsville. I couldn't help feeling like time had left us behind there in Odenville.

DISHEARTENING REALITIES

We survived the muggy warmth of summer, but as the weather cooled, we discovered there was no way to heat the house except for a small fireplace in the living room. We burned the coal slack my brothers scraped from an abandoned mine nearby and huddled together for warmth on the coldest days.

In our bedroom, my bed was placed next to the outside wall, while Lonnie and Johnny slept on bunk beds against the interior wall. During summer nights, we had enjoyed a cool breeze that filtered through cracks in the wall, but as winter approached, I shivered when the wind blew over my shoulders and across my face. To keep out the cold, I slept with a green woolen Army blanket over my head; it itched and made my skin blotchy. During the day, I saw light filter through the cracks, and sometimes I saw the moonlight at night, so I pasted newspaper on the wall to keep out the wind.

When Daddy saw my mess on the walls, he complained. "We look like poor folks with newsprint stuck on the wall like that."

"It's so cold you can see your breath like smoke," I said, blowing out to show him. "I put newspaper under our blankets to hold the heat and stuck some on the wall to cover the cracks."

The next day, Daddy brought home unused blue wallpaper left over from a remodeling job he had completed. My brothers and Daddy and I worked together to cut, glue, and hang the paper in our bedroom. Then we stood back and looked with pride at our handiwork. I asked, "Doesn't it seem like the room comes alive with the new color?"

Lonnie laughed. "It looks like the wallpaper is breathing in and out with the wind!" he exclaimed. We laughed with him, but we slept much warmer that night in spite of the wind whistling and the paper rustling.

As the days got even colder, Mother cried more, refusing to take her medicine because she didn't like the way it made her feel. Mother was beautiful and loving and kind to everyone until she had episodes of what she called a nervous breakdown. Sometimes she even heard people talking in her head. Her eyes went wild, and she talked about how much she despised Daddy. It made us sad and confused.

One night, Mother asked me, "Junie, don't you hear the voices? Listen, they sound like Oriental voices." I cocked my head and turned in every direction, trying to hear what she was talking about.

"What are they saying?" I asked, fearful for her.

She listened to the air, then said, "Can't you hear them? They're talking about war and bombs and shelters. We have to leave this house, run away and hide. Tell your brothers. Get your things now."

It was cold and miserable outside, and as much as I wanted a more modern house, I did not want another homeless adventure. I begged her to wait for Daddy and offered to pour her a cup of coffee.

She pulled a cigarette from the pack of Camels, placed it between her lips, and lit it with grand fanfare. Her long auburn hair fell across her forehead in a pretty wave. She pushed it back and took a deep puff, then blew out a swirl of smoke with a sigh, lifting the cigarette while crossing her shapely legs. Her gestures always reminded me of the pictures of Hollywood glamour photos.

The coffee and cigarette eased her mind until Daddy came home. The two of them talked in hushed whispers at first, and then Mother began screaming. "You buzzard gut old man! I'd rather go back to the hospital than take that medicine! It makes me feel awful!"

That cold and cloudy night, she opted to leave us and return to the hospital. As our parents drove away, my brothers and I cried. We knew what lay ahead—more work with no mother to help us cook and wash clothes and care for the baby, who was now nearly two years old. The worst part was not knowing what they did to help her at the hospital and realizing she could be gone for as long as six months or even a year.

My brothers knew she was gone again, but they didn't understand why. I tried to answer their questions, explaining that she couldn't cope with the worry of her brothers in the Korean War and perhaps the house and baby. I knew she heard voices that told her to do unimaginable things, but she refused to take her medication because it made her tongue swell and her mouth dry. She was miserable. I explained to the boys that some people had physical ailments in their bodies. Mother's sickness was inside her head. The hospital doctor insisted she would stay well if she took her medication and got plenty of sleep, but she refused to do that and had to go back.

Later that night, after I bathed Lee, put footy pajamas on him, and tucked him into his crib, it started to rain. Before I crawled into bed, I needed to write a speech for my civics class titled "What I Want to Be When I Grow Up." *That's easy enough*, I thought. I'd always wanted to be a teacher and a scientist who made discoveries or built rockets like they did in Huntsville. I fretted about how to begin my talk at school. Then I remembered reading Albert Schweitzer's autobiography.

Because the house was rowdy with three little brothers, I often studied or read my assignments while sitting in our pea-green faded 1949 Chevy. It was warm from the sunshine and blissfully quiet. Earlier that semester, I was to read an autobiography. I had chosen Schweitzer's *Out of My Life and Thought* because, as a scientist and medical missionary, he had won the Nobel Peace Prize. I was especially fascinated with how he worked in Africa to solve malaria by discovering where the mosquitoes laid their eggs in trays of stagnant water. I'll never forget his words: "Grow into your ideals so that life cannot rob you of them."[1]

After preparing for the speech, I left my papers on the kitchen table and went to bed. In what seemed like only minutes, Daddy hollered for us to wake up, asking me to get my brothers ready. He planned to ask our neighbor if she could keep baby Lee while the rest of us went to school. Dressing quickly, I woke my brothers, cooked oatmeal, tossed my papers into a ring binder, and then woke the baby. I hoped the neighbor, Lyda May, would come in time for me to catch the school bus that stopped in front of our house.

We stood on the porch and watched as the bright yellow bus slid to a muddy stop. I motioned for my brothers to run ahead and hop on. The bus driver honked the horn at me just as Lyda May drove up in her car, jumped out, and reached for the baby in my arms.

As I ran to the bus, the red mud stuck to the bottom of my scuffed brown leather shoes and squished up over the edges past the top buckles. When I hesitated to scrape the mud from my shoes at the bus door, the driver motioned me back, pulling the lever to close the door. As he shifted gears and lurched forward, I stumbled down the aisle to get a seat. Some of the kids whispered about our dirty shoes and our crazy mother. They moved and spread out so I couldn't sit next to them. Finally, near the back of the bus, I squeezed in with my brothers. In a singsong voice, older teenage girls dressed in pretty sweaters and poodle skirts sat on the back bench and taunted us: *The Kent kids stink. They got no kitchen sink.*

Ignoring them, we held on to each other, relieved that we actually made it to the bus on time. It was more than three miles to the school over slippery muddy ruts and potholes. It was still raining when we splashed up the walkway to the school building. Most of the girls got off the bus with umbrellas or hooded raincoats. My brothers and I raced to the front doors with uncovered heads. Before running to our classrooms, we tried to scrape the mud from our shoes on the big wire mats. When the bell rang, I skirted into assembly just in time to join the other students saying the Pledge of Allegiance.

Note

1. Albert Schweitzer, *Out of My Life and Thought* (New York: Holt, 1949).

Chapter 2

RAINSTORM LESSONS

A PROUD SPEECH

It was still raining as my class left our early morning assembly. I felt damp, and my drenched blond hair smelled like a wet dog, so I took a detour to the girls' rest room for paper towels. To my dismay, the same high school girls from the bus were there. One of them turned to me with a spray can of deodorant, held her nose, and swished the aerosol mist over me, laughing as she and her friends left the room.

Steeling myself, I went to first period math. At the beginning of the year, I had wanted to take algebra or geometry, but I was told those classes were for older students and usually only boys signed up for them. I knew I was smarter than some of the boys and even explained to them the math behind baseball batting averages and how many pounds of thrust it took for rockets to reach space. *Never mind,* I thought when I didn't get the more advanced classes, *I'll bide my time, and some day I'll take them.*

I completed my math assignment quickly, and then opened my binder to finish writing my speech for the next hour's civics class. I was prepared to say why I wanted to be a teacher and scientist, but I couldn't decide how to begin. I wondered if I should start out by saying something like, "From the moment the first person looked up to the heavens, we have dreamed of exploring and discovering unknown wonders." Or I could talk about tales in mythology such as Daedalus and Icarus, who flew too close to the sun on wings made of wax. Or I could tell how science-fiction authors like Isaac Asimov and Robert Heinlein influenced me, or how Jules Verne and

H. G. Wells inspired scientists and pioneers such as Robert Goddard to build liquid-fueled rockets or the Wright brothers to build airplanes. Or even begin with a story about Wernher von Braun who had a team of people building rockets to the north in Huntsville.

After working for several minutes, I finally felt ready. Moments later, I stood before the other students in my civics class, took a deep breath, and said with conviction, "I want to be a teacher to help students learn and maybe be a scientist and make discoveries." I told my classmates that reading Albert Schweitzer's autobiography inspired me to want to go to college to study science. Even though I heard snickers from the back of the room, I kept talking. I shared thoughts about the scientific method and about observation, trial, and error. I concluded by quoting an anonymous author: "To teach is to touch the future!"

Before I reached my desk, I heard chuckles and then whoops of laughter. The teacher, who had no control over the students, said nothing. One of my classmates jeered, "You're so poor, you'll never see the inside of a college, much less be a teacher." More uproarious laughter made my face flush with embarrassment. My heart was broken. *Those are fighting words*, I thought, holding back tears. Instead of letting them fall, I coped the way I usually did—by focusing on an object in the room (this time the colorful globe at the front) and daydreaming.

I spent the remaining minutes of class remembering some of my past experiences at more schools than I could count.

THE UPS AND DOWNS OF SCHOOL

First Grade: A Bright Beginning in Texas

Sadly, many of my school memories are like this one in Odenville—full of disdainful behavior on the part of other students and either silence or direct mistreatment on the part of my teachers.

My earliest memory of school, though, is a good one. I was fortunate to attend the same rural Irving, Texas, school in first grade for nearly a full year—quite a record considering we had packed up the car and moved a dozen times by the time I was eight years old.

My first-grade teacher, Ms. Gosset, loved her students equally, and we thrived on her enthusiasm, especially her sunny morning welcome, "Good morning, my delightful students! Are you ready for an exciting new adventure today?" Each week she used a different adjective to describe us until we learned the new word and could stand before the class to demonstrate it. To her, we were "wonderful," "amusing," "beautiful," "charming," or "lovely." If the student selected to demonstrate the new word was reluctant to use it, then he or she was allowed to act out the opposite or antonym. Usually, the boys chose to demonstrate the more unpleasant words like "bad," "nasty," "horrible," or other obnoxious adjectives that made us all laugh.

As soon as Ms. Gosset learned that I already knew how to read, she moved me rapidly through the levels of the *Dick and Jane* stories, then challenged me to read chapter books aloud to my classmates during story time. Instead of teaching everyone how to print the alphabet, she advanced some of us straight into cursive handwriting lessons. She taught us art and music and created learning games to challenge us. We all felt special because she identified every student's strength and honestly acknowledged our talents.

I loved everything about my first year in school; learning and making new friends was exciting. It was a grand and wonderful age of discovery, but the fun ended when I was not allowed to complete the first grade. Just after Easter, our family suddenly had to leave our home. Daddy told us, "Pack up what you can in a paper bag and get in the car. We're leaving tonight." Mother collected household necessities in a box to place on the floorboard of the back seat, then stuffed as much as she could fit into all the crevices. We left our toys and other belongings and drove from the outskirts of Dallas to Florida.

Second Grade: Darker Skies in Florida

In Florida, we moved from town to town while Daddy looked for a job. We might stay a week, then move on. We lived in the car, an old station wagon with a mattress laid in back where my brothers and I slept. In the first town—Orlando, I believe—I went to school in the second grade for only one week. Daddy returned to the roadside park where we waited after school and told us the sad news that there was no work. The plan was to keep driving south until we found a town that needed Daddy's carpenter skills.

By summer, we reached West Palm Beach. Daddy got a job, so we rented a house and I was enrolled in the second grade in yet another school. The teacher insisted that every one of her students should learn to write in cursive, but I had already learned to do that in first grade. I'd never learned how to print block letters. The teacher, though very strict, made the writing and spelling lessons fun.

At yet another school, we lived close enough in our rented house that I could walk to and from school. In this second grade class, we were given many art lessons. I remember that after we finished our assignments, we were allowed to sketch and paint pictures. I watched other kids draw pictures of airplanes and boats or bicycles and horses.

I wanted to draw a picture of my dream house. Every day I drew the same picture, often adding a new feature—a chimney with smoke, then two windows with curtains and a lamp placed on either side of the front door. As days went on, I added a tree growing out to the side with a swing hanging from the branches that reminded me of our first pretty home. Daddy taught me how to draw a bird and suggested that I draw out the floor plan that matched the shape of the house. Eventually, I drew a straight line of tulips across the front of the house.

I liked that school, but we had to leave the rental house and neighborhood because my brother and I got worms in our feet from the sandy dirt we played in. When white and red streaks reached our legs from the sores, Mother wanted us out of that filth, so we had to move again and leave behind the friends I'd made in school.

Daddy told us, "Well, my little family, I was hoping to wait till Christmas to surprise you with a very special gift that's not quite ready yet, but since your mother wants to move, it's time to show you your early present." We packed up our grocery bags of stuff once again and drove to the other side of town not far from the beach where Daddy had bought land to build us a house with the money he earned from working at his steady job. The wood-framed house was rather small, and though it wasn't finished on the inside, it was going to be our house.

As we jumped out of the car to explore, I gushed, "It's beautiful! This will be our very happy home forever!" I remember that we could walk from room to room between the boards of the walls that held up the ceiling. We didn't have much furniture, but to us kids it felt like camping out.

By spring, our parents were beginning to argue more and more. Mother, would yell at Daddy, "You're just an old buzzard gut! When are you going to finish these walls and hang doors so we can have privacy?"

Daddy returned the insults. "You ole biddy! You nagging slave driver! I work all day, then come home to walk behind a mule so we can get a garden in before the rain comes."

The situation between Mother and Daddy didn't improve, and one day in February while Daddy was at work, our mother packed up my two little brothers and me and walked miles to the highway, where we stopped a greyhound bus to take us back to Texas. I pleaded at every step, "Why can't we take Daddy with us?"

Mother promised that we would return to Daddy some day. The bus stations were dark and scary, and some nights, we slept on benches and the floor waiting to make connections. We lived in these stations or on the streets until we finally arrived in San Antonio, where we lived with grandparents for a short time. I finished second grade there. In each new school, I had learned to make friends with the girl who smiled the most. We moved several more times, and Lonnie and I even spent some time in a foster home. Eventually, Mother got on welfare and then got a job so we could rent an apartment at Victoria Courts, which was subsidized government housing. Johnny, the youngest at the time, stayed much of the time with our grandparents, but he was with us on most holidays and some weekends.

Second Grade Again: Texas

For reasons I didn't understand, I had to begin second grade again the following year. I was placed in the learning-disabled class where I encountered children who growled rather than talked, pushed and shoved for entertainment, or hid under desks eating fists full of white paste. I remember a boy who sat next to me and always smelled like urine.

Our teacher's fierce glare terrified me so much that for the first week of school, I froze with fear when she approached me. She pulled her hair, black as coffee except for a grayish yellow streak above her left eyebrow, straight back into a bun at the nape of her neck. Her voice crackled, and her eyeglasses often slid down her pointy nose. Her pale, thin body appeared ghostly in the black dress that touched the tops of her leather shoes. She carried a thick ruler that she used to smack unruly children on the knuckles. When students were unusually rowdy, she hit their arms or legs with a plank-sized paddle; on occasion, she also laid children across her lap and spanked them.

Most of the time, she sat at her desk, ordering children to come to her. We were instructed to call her "Mam." On the first day that I was brave enough to raise my hand and ask a question, I said, "Miss Mam, I finished counting the stick shapes in my workbook and marking the answers. May I draw a picture on some paper?"

"The name is just Mam," she said, "not Miss Mam, and no, you may not color. You are to sit at your table, hands folded in your lap with your smart mouth closed, and wait for the rest of the class to finish the lesson."

I looked around the room at kids scuffling, sleeping, and picking at noses, and with my idle time I chose to daydream. I noticed that the cracked and peeling walls needed paint, and the dusty chalkboard could use a wipe. The date was written in chalk in the top right-hand corner. It was close to Halloween, and the year was 1951. Looking up, I noticed dim lights on fuzzy electric cords hanging from the ceiling. The wooden floors were covered in what looked like the dirty oil my dad drained out of his car. A faded map of the United States hung torn and ragged near the teacher's desk, reminding me of my family's dozens of cross-country trips.

Daydreaming helped, but it couldn't effectively remove me from the horror of Mam's classroom. In all the schools I had attended, I had never seen such mistreated and unruly children. I was so frightened for us that my stomach hurt and I couldn't sleep at night. One day, as I cowered in her class, the pit in my belly threatened to explode into tears. Suddenly, I heard a dripping sound coming from the boy who sat in the seat next to me at our table. The odor was familiar. I turned to look under his chair and saw a giant puddle of urine.

The dam holding my tears broke, and I started to sob. "Mam, I want to move my chair to a different table!"

Mam stomped toward me and barked, "What's wrong with you? It's just pee. Go get the mop out of the cloak room!"

I did as I was told, still crying. The other students laughed, but a husky bully of a boy thought I had been the culprit and teased, "Little pee-pee baby's crying."

I sobbed until Mam sent me to the office for shouting and disturbing the classroom. It was not the first time I had been sent to the office, so I knew the way. I was relieved to be out of that smelly room. When I reached the principal's office, I tugged on the heavy door and pulled on the handle high above my head. Entering, I climbed on the big sofa and waited for the principal. I expected my usual paddling and then a return to the scene of the "crime," so when the principal stepped in front of me, I spewed out one indignant point after another: "Why did you put me into that classroom? I'm supposed to be in third grade. The teacher doesn't like me and the kids don't like me either, and anyway, I know how to do everything already. It's just stupid! Can't you do something?"

The soft-spoken principal grimaced and said, "Well now, young lady, let's see who you are." Checking a folder, he said, "Hmmm, oh, yes. When you and your mother came to enroll you in this school, she had only one report card that showed you had spotty attendance last year, and when we asked you to print your name on a form, you said you couldn't print."

"No," I argued loudly. "I can't print well because I learned to write cursive in first grade. I had a good teacher." I explained how Ms. Gosset had made games out of learning, leading us through store and bank simulations to teach us about money and social interactions. I told the principal how I read chapter books in first grade, enjoyed art and music, and played outside after lunch every day. Then I jumped off the sofa and shouted, "I hate this school! I don't ever want to go back to that classroom!"

The principal pulled out a chair and sat at my level, quizzed me more about my education, and finally ordered me to return my precocious little self to the classroom. That night I asked my mother what *pah-ko-shus* meant and told her what had happened. Too tired to listen, she said, "Junie, he probably meant *obnoxious*," and then turned to go to bed.

At only nine years of age, I thought, *Growing up is not easy. It's confusing and frightening.* I remember that year clearly for all the trauma, but also for the concern that nobody loved or even cared about us. I frequently stared out my window to the starry night sky expecting some kind of response from our "Father in Heaven," whom I imagined sat on a big throne and answered prayers. I pleaded for God to help Daddy find us.

I was beginning to understand why I was put back a year in school and enrolled in a classroom of learning-disabled and problem children who fought, screamed, smelled bad, bullied me, and made puddles on the floor, but knowing this did not help. I still cried every day before, during, and after school.

A School Assembly Challenge

Later, during a school assembly, the students were told that anyone who could memorize the entire chapter of 1 Corinthians 13 in the Bible would be able to recite the chapter in the school Christmas program. I saw the challenge as an opportunity to prove my intelligence and get moved to another class. Every afternoon after doing my chores, I used my child-sized New Testament Bible and memorized the chapter. It was all about faith, hope, and charity or love. Even at that early age, I focused on hope. You can decide how much or how little faith you want, and you can decide on the depths of your love, but hope is different. In my life at that point, memorizing that chapter held out a small glimmer of hope for me that the school staff would rescue me from Mam.

I worked hard to remember every word, but ultimately Mam refused to allow me to recite before the assembly. "Sit down! Be quiet!" she hissed.

With my hope crushed once more, I frequently got in trouble with Mam. Was it because I was precocious? Actually, Mother was probably right. I was obnoxious, and I was also devastated and miserable. I was scolded, even paddled, daily. I just wanted to escape, even if it meant being sent to the principal's office.

The Gift that Made the Difference

During the Christmas holidays, someone from Mother's office gave her a book and tangerines. I was delighted that she accepted them, as Daddy was always too proud to take charity as long as he could put food on the table.

The fruit was luscious, sweet, and juicy. I sat at the table peeling a tangerine and noticed the name of the author on the book: Peale.

Suddenly, I began turning the pages of Norman Vincent Peale's *The Power of Positive Thinking,* reading every word. One of the book's strongest messages was that positive thinking can help a person cope with even the worst of circumstances, and that God is always with us. Soon, I was more absorbed in the book than in the tangerine. More than anything, I wanted to get out of Mam's class. I was determined to think or pray my way out.

To make the book more practical for a girl my age, I created a lesson—the ABCs: A Guide for Positive Thinking.

A was for attitude. I needed to change my attitude and view my problems as challenges, forgiving others along the way.

B was for belief. I needed to believe in myself and in the Power greater than myself.

C was for commitment. I needed courage to create a positive out of a negative.

Later, I added D for dreams.

First, I chose to change my attitude about being stuck in a difficult classroom. I told myself that the teacher, the other kids, and I were all stuck there together.

Next, I believed that God could help me with my problem, so I prayed, "Please forgive me for my mean and bad thoughts. Give me courage to talk to Mam so I can help the other kids." I decided to try to speak with Mam privately rather than making demands or shrugging her off.

Finally, I committed to begin helping Mam with the children, reading to them, caring for them, teaching them their numbers and songs, and striving to ignore the cruel words some of them spoke.

One day not long after my change of heart, I was rewarded with a gesture of thanks from Mam. I finally realized that she felt bitter about being forced to teach a class in which she was ill prepared to work with learning-disabled students. Her reaction to her lack of self-confidence translated into misery and cruelty. After a while, I noticed that she put away the tools of corporal punishment and actually grew a smile.

A Pivotal Test

One day, a stout man in a dark suit came into our classroom to make an announcement. With his deep but hollow voice, he spoke through a fuzzy, food-stained beard. "Now children, we are all going to march in a straight line, without talking, to the school cafeteria to take a test."

"Why?" I asked. "What kind of test? What's it for?"

The man's nose and cheeks turned beet red. He answered gruffly, "According to how you score on the test, you may possibly be placed in a different classroom."

Remembering my ABC guide, I chose to believe I could read and answer those questions, and I believed God would help me. It was exhilarating to work through the test with confidence and solve the problems.

When the bearded man returned to our classroom a few days later, he asked, "Who is June Kent?"

Once again, I was taken to the principal's office. I sat on the giant sofa and waited anxiously with my hands folded in my lap. Suddenly, I looked up to see six adults march into the room and stand around me.

I barely recognized the principal. For the first time, he was grinning with a somewhat shocked look in his eyes. He spoke first. "So, it's you!" Then he asked, "You remember the I.Q. test you took recently?"

I nodded.

"Well, young lady," he continued, "you made the highest score of any student in the entire school."

They all stared at me in silence, waiting for a reaction. I blushed at the attention, cleared my throat, and asked, "Does that mean I can go to a different class?"

They chuckled, but one of them took me by the hand and escorted me to a newer building with shiny wooden floors. The classroom looked friendlier, with activity stations placed around the room. My new fourth grade teacher welcomed me. She introduced me to the students, who looked up with smiling faces and cheerfully said, "Hello!" Sunshine filtered through the windows. All the students were busy, no one was eating glue or crying, and the room smelled like the roses in the vase on the teacher's desk. I felt happy and sad at the same time. Though I was thankful to be in my proper grade level, I regretted leaving Mam's classroom now that I knew I could help the other children.

That day, I used my ABC guide again and committed to create a positive out of a negative. I decided I would grow up and be a teacher, not an angry one with no patience, but a good teacher who loved and cared about her kids. I would help children and give them opportunities. At that point, I added "D" to my guide. I would be a dream maker to offer hope and help children make their dreams come true.

A PROUD COMMITMENT

It was years later that I sat in shame after giving my proud speech in civics class. As the bell rang to go to the next period, I collected my books and papers. My daydreams had reminded me of the difficulty of those early school years, but I also recognized that I'd learned much more than the lessons teachers taught. And I remembered my commitment to help make dreams come true.

Somehow, at age thirteen, I was still struggling to prove myself. I knew I had to go back to my ABC guide. Quickly, I whispered a prayer. "God, please help me forgive those who said hurtful words, and make me confident so I can laugh at myself more easily."

Right then, I made a commitment to my future. As I walked down the stairs heading toward my next class, I said boldly to myself, "I *will* be a teacher someday! And I just might come back to teach in this school!" Tossing my books into the locker and slamming the door closed, I shouted to myself with conviction, "Oh, yes! I will be a school teacher . . . and maybe even a professor or a scientist!" Then I ran out the double doors of the school building to the gym for some heavy-duty exercise during PE class. It was just what I needed.

Chapter 3

THE DAWN OF FAITH

SPRINGTIME ON THE FARM

Spring came to Odenville in late April 1957. It followed the coldest winter I had ever known. Rain was more often sleet, chilling us to our bones. The contrast in the spring weather was noticeable everywhere. Sun streaked from behind wispy clouds in a pale blue sky. Birds chirped, building nests in the tree branches. Green blanketed the fields and meadows again, and violet and yellow wildflowers appeared all around the house.

Spring introduced new farm chores. On either side of our faded red house, Daddy planted a vegetable garden. The produce grew quickly, and my brothers and I took turns hoeing the weeds out of the rows of tomatoes, green peas, cabbages, and corn. Our family enjoyed the results of our labor at mealtimes. Our Guernsey cow birthed a calf one afternoon right in front of us. The mother licked her child clean, and soon the calf put out his back legs, then front legs, and wobbled up on all fours. He nuzzled his mother and started to nurse. Sights, sounds, and surprises like these made our work exciting.

For fun, we went skating on Saturday nights at the roller rink in the next town, driven there by a neighbor named Billy. Billy was sixteen and had dropped out of school, gotten a job, and bought a car. It was fun to put on the skates and circle around the rink to the music.

Later that spring, a drill truck backed behind our house, and for days it drilled and pounded into the earth to dig out a well. The simple job grew complicated because the workers hit a coal mine shaft. After putting a metal sleeve in place, they kept drilling until they finally found water, removed the truck, and helped Daddy build a small well house. We could not afford an electric pump, so we had to turn a crank and lower a bucket down into the well, and then pull it up full of fresh water.

Daddy also bought a butane tank and a kitchen stove. On cold spring mornings, I stood in front of the open oven door to get dressed. These upgrades made life on the farm a bit easier.

LOVING, LEARNING, AND LOSING

We lived in Odenville for two years. During our second year, Mother went back to the mental hospital. Our neighbor Lyda May, mother of our friend Pee Wee, took care of the baby, now a toddling two-year-old, during the day, and I took care of the family after school. With Daddy's help, I cooked dinner at night. Life without Mother was like a nightmare from which I couldn't wake. I was mad that Mother left us again, and when we were cold, I jealously imagined her in a warm building with a real bathroom inside the house.

It was difficult to balance being a young girl and being a caretaker. My hopes and dreams always seemed to take last place. At school, the counselor discouraged girls from taking advanced math classes, and instead persuaded me to learn how to sew in a home economics class. Once I learned to sew, I begged Daddy for a sewing machine. "With the animal feed sacks, I can use a pattern and make skirts for me and shirts for the boys," I explained. Within a few days, Daddy brought home a new, portable Singer sewing machine with its own case. I enjoyed the creative opportunity to sew, learning how to make pockets on my brothers' shirts and play clothes for the baby.

One day, two men dressed in dark suits came to the back of the house and stood at the rickety stairs talking to Daddy. I watched as he kept shaking his head. Finally and reluctantly, Daddy stepped inside, gathered up the portable sewing machine and the box of parts that went with it, and carried them out to the men. The machine was "repossessed," a new word for me. It meant Daddy had not made payments, so they took the purchase away from us. It was a terrible sign to me of the hopelessness of our situation. I remember crying out, "Why are we so poor? Why can't we have anything? Why do we have to live like this?"

A TURNING POINT IN FAITH

After the men drove away with our beloved Singer, I jumped off the porch, ran past the garden, and hurried into the woods near our home, weeping over the injustice. The path was strewn with pine needles and overgrown with thorny bushes, and I stumbled and fell. For a long time, I lay there pounding the ground. Then I threw fists of dirt and shouted obscenities no one could hear. Finally, I collected my wits, felt ashamed for losing my temper, and sat in silence, pouting about the state of my life. After a time, I prayed, "Hey God, remember me? It's June. I'm down here in these woods, and I don't know what to do, how to help my family, where to start." Sobbing again, I asked, "Can't you help me?"

Suddenly, I recalled my ABC guide to positive thinking. How could it help me now?

A—accept who I am and where I live. I thought about the positives: some aspects of school were good, even though I was discouraged from certain classes.

B—believe in myself and in God. I knew that if our family could overcome adversity when we lived alone with Mother, then surely we could find ways to solve our problems with Daddy. I needed to remember the song I taught the children in Sunday school: "If God so loves the little birds, I know he loves me too."

C—have courage to make a commitment. Standing up and dusting the dirt from my hands and clothes, I committed to try to make a difference and work to solve our problems.

As I slowly ambled along the path back home, I made a promise. "When I grow up, I will get a job with a steady income, keep a budget, pay my bills, and never let this sort of thing happen to me." I knew I could not solve our immediate family problems alone, but I asked God for help. That moment stands out as a spiritual turning point; I was certain that God would guide me through the present storm. Unexpected gratitude filled my heart for the mighty Power whom I trusted to help my family and me.

At home, I resolved to make a difference, but first I had to complete my homework. Determined to find ways around my obstacles, I continued to sew with a needle and thread. I cut the fabric from a pattern, and if I made tiny stitches, the seams held together. Seeing my raw and pricked fingers, my home economics teacher gave me a thimble, and others began to help too.

In a 4-H meeting, I learned about hatching baby chicks and knew this could help our family. Daddy agreed and got me several dozen fertilized eggs.

Using boxes surrounded with light bulbs, we built an incubator in an old shack across our dusty dirt road, and watched the baby chicks hatch one after another. We fed them and cared for them, and when they grew up, I begged not to have to kill them, pluck their feathers, and sell them, so every week Daddy took a crate full and sold them at a market in the city. We made enough money to buy jeans for my brothers and new shoes and warm socks for the upcoming winter months.

Later, I remembered those baby chicks hatching and wondered about my turning point with God. Does a growing soul need to incubate like a chick? Does it need special nourishment? When the soul is ready, does the person break out of her old identity to discover a new self? Would I always consider myself new now that I recognized I had a soul?

MOTHER'S RETURN AND NEW FRIENDS

In late spring, Daddy and I drove to Tuscaloosa hoping to bring Mother home from the hospital. On the way, Daddy surprised me by driving through the campus of the University of Alabama. I knew college was a place for advanced study after high school, and I wanted to go there to learn to be a teacher.

Daddy pointed to young women riding in a convertible and said, "Junie, that could be you someday. Look at all the fun they're having!" I studied them longingly, then asked dozens of questions about how we could make it work. It sounded hopeless for our family, but Daddy encouraged me. "Don't lose the faith," he said. "Miracles happen, you know. Just keep up your studies and continue to keep your nose clean." I wondered what my nose had to do with it, but I didn't question his rare words of wisdom.

We were able to get Mother released, and she came home with us that day, happy and grateful to be back with her family. She smiled and hugged us and sang songs while she busied herself in the kitchen or read books to the baby. It was wonderful to be a family again. I felt the weight of responsibility roll from my shoulders and experienced a newfound joy over having a mother in my life again. We quickly resumed a relationship.

One late afternoon as Mother and I worked together in the kitchen, I told her how much our neighbor, Lyda May, had helped us with the baby. I also mentioned my frustrations in school.

Curiously, Mother asked, "What's bothering you the most about school?"

I wanted to tell her about the teasing, but instead I explained, "I don't have any friends. We've moved so much all these years. I guess I never learned how to make real friends."

She wanted to know what I would talk about with a close friend, and I said, "Important things like rockets they're building in Huntsville, or how airplanes fly, and scientific discoveries that could cure cancer."

Mother paused, then said thoughtfully, "You made friends in Birmingham just before we moved. Remember, they come in different sizes, shapes, and interests of their own. Why don't you try being a friend to them and learn what their interests are?"

I thought her advice sounded doable. Most of all, I treasured having another woman to talk to about life again.

Over the next several days, I did manage to make a few friends. Eugene liked baseball, so he and I talked about the Birmingham Barons and New York Yankees. He was a smart student, I thought, and nice enough to give me some packages of chewing gum one day. I giggled nervously at his attention. He was funny too, always joking with me. Often when we were going separate ways, he'd say something like, "Hey, June, what did the baseball glove say to the baseball? Catch you later!"

A studious and clever girl named Jessie also befriended me. We worked together on homework assignments and talked about television shows she watched, like the *Ed Sullivan Show*. She told me about how Elvis gyrated across the stage, and how her parents thought it was disgraceful.

Johnny was another smart student. He laughed often with other students about Saturday night television shows. Of course, I couldn't relate because my family didn't have a TV, but I laughed with them anyway. Eventually, I learned that Johnny was also interested in rocket science and physics, the subjects boys were encouraged to focus on in later years.

When the bus dropped us off from school, my brothers and I bounded into our front door, feeling grateful to have our mother again and to see our family all together. I smelled the lingering freshness of a clean house, the fragrance from furniture polish in the living room, and savory aromas coming from the kitchen. We all talked at once, telling her about our day and what we'd learned.

Life was good again. Within days, Daddy got enough money to hook a pump to the well and pipe water into the house. He fitted a large sink into a cabinet and placed a faucet above it. We had plenty of milk, eggs, and a garden filled with seasonal vegetables. We learned to can corn and peas and put up peaches for the winter. Our goat ate the grass when it grew tall in the backyard.

As I lived these good days with my family, I thought about the faith that got us through the bad times and helped us celebrate now. During our two years in Odenville, I learned to appreciate both of my parents and their sacrifice to care for our home and family.

Sometimes people called me a "Pollyanna," a dreamer, and a wishful thinker. I met many negative people along the way, but my faith changed my heart, giving me hope and teaching me to believe in myself and in the mighty power of God. It also helped me focus on my goals to study and make good grades in school, to have courage to turn negatives into positives, and to commit to work hard to see my dreams come true. Faith, hope, and love served as my inner compass to overcome adversity.

Chapter 4

FIRST STAR WISHES

AN UNEXPECTED MOVE

In 1957, on the last day of tenth grade in Odenville, my brothers and I rode the bus home, excited about the freedom of the summer months. Lonnie and Johnny took turns talking about swimming, hiking, and wishing on first stars.

"Ah, the lazy days of summer," one of the older girls swooned, stretching her arms up toward the top of the bus.

Lazy days? Wonder what that's like? I thought. With all the farm chores, I understood freedom from school and homework, but not laziness.

When the bus stopped at our house, a group of kids (the ones who had never given up mocking us) clapped and yelled, not because we were getting off the bus, but because they saw a homemade sign planted in the middle of our front yard. Painted in bright red fresh paint were the words "For Sale."

Surprised beyond belief, we clambered off the bus and raced across the yard, up the stairs, and through the front door into the house where both of our parents waited for us.

"What? Where? How? Why?" The single words fell out of our mouths in unison.

Daddy calmly suggested that we sit down and talk about our family's future.

Mother began, "You recall how cold last winter was? And Junie, you want to learn more about rockets and space satellites, and boys, you've asked to go swimming in the ocean."

"So," Daddy chimed in, "we're moving back to West Palm Beach, where I've heard I can get a good job because they're building houses in subdivisions. It's just south of that place where the Air Force is launching rockets out on some cape."

Before we could express our sorrow or interest, Mother concluded, "I've put a grocery bag on each of your beds. Spend the rest of the day gathering what you want to take that will fit in the bag."

At that, I expressed my concerns. "We can't leave all this behind for another move! We have furniture and dishes and pretty things we've collected, and Daddy built some great things for this house. Can't we put stuff in the truck to take with us?"

Daddy explained that he wanted to get us settled in our new house in Florida. Once he did that, he'd return to Odenville to sell the old house and pack the rest of our things into his pickup truck. He looked at my dazed eyes and stunned expression and teased, "I'm pretty sure we can find a house with three bedrooms and at least one bathroom to buy or rent."

Mother and Daddy packed the car as tightly as they could, barely leaving enough room for four children and our small paper grocery sacks. I filled my bag with a few summer clothes, a swimsuit I had outgrown, and my child-sized Bible and favorite book. I noticed that Mother had packed her treasured rose and sage green bedspread. It was then that I knew for certain we were on our way toward another family adventure or disaster.

The next morning, we left Odenville, where we had lived for two years, the longest we had ever lived in any one place. With mixed emotions about leaving, I reflected on our experiences. We'd had many disasters, tribulations, and sorrows, but we'd learned lesson from all of them. Most important, I'd learned about the kind of person I wanted to be, how to make good and lasting friends, and how to bond with my family members and work to accomplish what seemed impossible.

ON THE ROAD TO FLORIDA

As we traveled south toward Florida, the sun rose higher, and the heat surrounding the six of us grew stifling. Even with the windows lowered, the hot, dry air only helped evaporate some of the sweat beads trickling down my neck and across my back, which was stuck to the plastic seat covers. When Daddy stopped for gas, he gave me a dollar to buy lunch at the adjacent grocery store.

Lonnie and I walked the aisles of the country store and studied the prices. We found a loaf of bread for 20 cents, a quart of milk for 30 cents, and a pound of bologna for 50 cents, which added up to the dollar. Then Lonnie suggested putting the milk back since it wasn't enough for all of us. He reasoned that we could buy six nickel candy bars with the rest of the money. I agreed. Most of our traveling meals were that way, with the occasional jar of peanut butter, pound of bacon, or dozen eggs thrown in (Mother cooked in a cast-iron skillet over a fire at a roadside rest stop).

When we reached the coast of Florida's Gulf of Mexico, Daddy pulled off the highway and parked on the sandy shoulder so we could see the ocean. Tired and hot from the long ride, we all piled out of the car and rushed to put our feet into the salty water. Whooping and yelping, we got our clothes wet to our waists. Giving up, our parents said, "Go on and jump the waves. We'll stay here on the beach for the night."

Later, we collected driftwood, dug a pit to make a fire for roasting wieners, and laughed and told stories on each other. Then, wrapped in towels Mother retrieved from her box in the car, we listened to the soothing sounds of the sea.

STARLIGHT MEMORIES

The fire dimmed into red embers, so we lay back on our towels and gazed at the sky. I thought about the wonder of their creation and recalled how my "first star wishes" to have our family together came true. I told everyone how to find the North Star and pointed out the constellations I recognized, explaining facts about them as I went.

Eventually, my brothers groaned and complained that they were tired, but Daddy asked, "Did that Odenville school teach you about all those stars?"

Cautiously, I shook my head and told him I learned about them in a book I had "borrowed" back when Lonnie and I lived in the foster home after Mother made us run away to San Antonio. I was ten years old at the time. (Our grandparents had kept Johnny.)

As Mother, Johnny, and Lee slept cuddled together on the beach, Daddy, Lonnie, and I sat and talked about my self-education. In the foster home, dubbed a "boarding school," children were not actually taught until they reached high school age. The teacher forbade me to join the older kids and said I couldn't use the library for several more years.

One day when everybody left the room for lunch, I had slipped into the library and looked at all the books. My eyes were drawn to a shelf with a row of beautifully bound volumes—a set of the Encyclopedia Americana. I pulled down the first one, bound in a maroon cover with gold letters. Opening it like it was a treasure box, I nervously thumbed through the pages. My eyes fell on the words *aeronautics*, *anatomy*, and *astronomy*. I wanted to read it all. Without even a second thought, I told Daddy, I had run upstairs and hidden the book under my bed.

I explained that I had lain under my bed with a flashlight reading the "A" encyclopedia book. If the teacher, Ms. Young, stepped into our room and switched on the light, I scampered from under the bed and kneeled as though I were praying. Turning out the light, she would grunt, "Jesus, she prays all the time!" Even with the danger of being caught, I built up quite a bit of knowledge about the subjects that mattered most to me.

Telling Daddy about our time in the foster home subdued Lonnie and me. We remembered Mother calling sometimes to tell us we could not come home on our one free weekend a month. Sitting by the ocean that night, I could still remember how much that hurt. I remembered the bullies Lonnie and I had encountered and the depth of longing we had for home.

"Daddy," I asked, "why is it good to forgive but so hard to forget?"

He thought for a minute, then answered seriously, "Until we truly forgive, we're imprisoned and locked in the memory. Forgiving gives us freedom to open the door to the future and see the wonder of possibilities, like those stars." He gestured to the sky, brilliant with more stars than I'd ever seen before. Looking up, I too could imagine the possibilities of the future.

The morning after that thoughtful night, we woke with sand in our ears and hair. It even seeped inside our clothes, making us itchy and sticky. We shook off as much sand as we could, then climbed inside the car expecting a miserable ride. Soon, though, Daddy stopped at a gas station, where we used a water hose to wash the sand from our bodies so we could continue the drive in relative comfort.

The drive across the panhandle of Florida to Highway 1 took another day, but the reward for me was the drive past Cape Canaveral. As we drove south, Daddy pointed it out. I imagined highly creative and educated people building rockets and satellites and missiles and setting them up to launch.

Thinking out loud, I said, "Don't you think it's magical that we lived near Huntsville where they built the rockets, and now we're moving near the Cape where they will launch them?" I couldn't believe it.

My brothers took jabs at me, both giggling. "You're weird! What's magic got to do with it? You're a fruitcake—a space nut!"

I laughed at the truth of it all. I was hopeless, but for fun, I said, "Oh, yeah? Well a big oak tree today was a stubborn nut yesterday that held its ground and took root." I wanted to live somewhere long enough to grow roots.

FIFTEEN IN FLORIDA

The summer we moved to West Palm Beach, I quietly celebrated my fifteenth birthday. We moved into a three-bedroom apartment that had once served as military barracks. My brothers and I swam in the canal behind the rental property until we saw alligators in the water with us. Our neighbors on one side had moved from New York. Their daughter Kathie was my age, and we became friends. On the other side of our home lived a policeman and his family. They acted rather distant, but I still liked them and babysat for the children. One morning while I watched them, the kids made toast for breakfast. They piled on gobs of butter, then peanut butter, then jelly. I'd never seen that much food put on one slice of bread.

Daddy returned to Odenville with the pickup truck, but he brought back only a few things from our home. He said he had sold the house and most everything else, and with the money we could go to the Farmer's Market to buy used furniture and even a used television. The day he came home with the TV, he also brought me a used record player and two Elvis records. It was my best birthday ever.

A PROMISING NEW SCHOOL

The week following my birthday, we all registered for school. My brothers attended a neighborhood elementary school, and I caught the bus at the end of our street that took me all the way into the city to Palm Beach High School. I'd never seen so many students. The campus was as big as the university campus in Alabama. While my school in Odenville served about two hundred kids, this high school had thousands of students. They drove up in cars, walked across campus in beautiful clothes, and greeted each other like family. It was exciting but terribly intimidating.

The new students lined up to talk with a counselor and register for classes. When it was my turn to fill out the forms, I asked the counselor, "Do you allow girls to take chemistry and physics and algebra and trig?"

"Of course," she said, looking puzzled. She instructed me to write down the classes I wanted to take, and said I would receive a schedule with the information I needed to get to them. Then she told me to explore the campus and eat lunch before returning.

I walked around the campus in a daze. Finally, I would be allowed to study the subjects I had always hoped to learn, but would I ever learn my way around that big place? Eventually, I found the cafeteria. I stepped into a line of kids, picked up a tray, and looked to see what I could buy that cost a quarter. The huge variety of choices confused me, so I simply bought a carton of milk. I found a table, poked a straw into the carton, and sat next to a group of girls who ignored me. I looked at my shabby clothes from Odenville, then at them in their cute skirts and sweaters, and decided to do something about my clothes.

When I went back to collect my schedule, the counselor told me I would have to take other classes to build up to some of the ones I'd chosen. I didn't mind. I liked going to school. The teachers were passionate about the subjects they taught and inspired me further to be a teacher. I settled in quickly and even made friends. Janice had a car and a stack of record albums as tall as she was. She invited me to meet her for lunch at a snack bar and soda fountain off campus across the street. Kids selected songs on a jukebox, drank cola with cherry flavor from the soda fountain, and ate hamburgers. I'd never seen anything like it.

I also made friends with my lab partner in chemistry. Like me, he was interested in rockets and the space program, and in late September, he asked me to go on a date with him. I was so surprised that I didn't know how to respond. I was insecure and afraid to be alone with a boy, so I asked that we get to know each other better first. Janice suggested that I make it a double date. I thought about it for a long time, recalling what one of my Odenville teachers had told a group of us girls: "Never kiss a boy unless you are willing to marry him because you never know where that kiss will take you." It was fairly practical advice.

THE SPACE RACE HEATS UP

On October 4, 1957, everything changed for those of us who took science classes at school. After that day, our discussions moved from the usual daily routine into concerns for our nation. The USSR had blindsided the United States with the launch of *Sputnik*, the first man-made Earth-orbiting satellite. It merely circled the Earth about every ninety minutes, sending out a beeping radio signal that, according to the nightly news, was "a technological Pearl Harbor that shocked the world." Dismayed, we huddled in groups. I was furious. After all, Americans were supposed to launch a satellite that year. I couldn't believe the Russians had beaten us to it.

That night and on the days that followed, Mother became increasingly agitated while we watched the news. "What are those Russian Communists doing? Are they using cameras to spy on us? Why does it make a beeping sound? Is it going to drop an atomic bomb?"

The more interested I grew in the news, the more frightening it was for Mother. Finally, Daddy and I stopped talking about *Sputnik* and tried to keep all news away from her. Then, a month later, the news exploded with word that the Soviets had sent up *Sputnik 2*, which carried a dog named Laika, who died when her oxygen ran out.

A LONELY CHRISTMAS

Mother's nerves couldn't handle it, and she told us she needed rest. For days she hid in her bedroom. When she finally came out, she said she wanted us to go with her back to San Antonio. We knew she needed help and medicine, but we were unfamiliar with the clinics and hospitals in Florida.

Soon afterward, as I walked home from the school bus stop, our police officer neighbor stopped me to ask if my mother had been in mental institutions. I told him she had nervous breakdowns, but assured him she never caused harm to anyone and certainly would not hurt her family or neighbors. We simply needed to find a doctor to renew her prescriptions. I suggested he talk to Daddy.

The next day, while we were at school and Daddy was working, the police came and took Mother away, first to the city jail and then down to a hospital in Miami. My friend Kathie's mother, who kept the baby, told us what had happened.

"The police took her to jail?" I cried with indignation. "She's not a criminal. She just needs her medication." I burst into tears.

When Daddy tried to find Mother, he learned that she was committed to the mental institution in Miami and that we could not see her for weeks, maybe months. I was beyond consoling and blamed myself because of what I'd told the policeman.

After that, I had two lives—one at school where I happily studied the subjects I loved, and one at home with the renewed responsibility of caring for a family. I failed at making a loving home for them. I tried, but I couldn't even create a pleasant Christmas experience. Daddy gave me twenty dollars to buy gifts for my brothers, and all it covered were a few small toys and books.

Daddy brought home a tiny Christmas tree that we set up on a table to make it look bigger. We decorated it with homemade paper ornaments and a popcorn string. It was unseasonably warm in Florida, but it seemed strange to wear summer clothes during the holidays. Most of all, we were sad without our mother.

ENTERING THE SPACE AGE

Amid these trying times at home, finally, on the night of January 31, 1958, America's first satellite was successfully launched. *Explorer 1* soared into space on top of a version of the Redstone rocket known as the Jupiter C. I watched the sky to the north over Cape Canaveral, and Daddy watched television late into the night, waiting for an update. Reluctantly, I followed everyone else to bed, but I rose at the crack of dawn the next morning to watch news of the launch. There, before my eyes, I saw the rocket lift off the pad atop a great plume of fire and smoke. The United States blasted into the Space Age. Waving my arms, I shouted, "It worked this time! We've finally entered the space race!"

Daddy ran into the room and watched the film again with me. He shook his head and cautioned, "But, my little space bug, the United States has a lot of catching up to do."

The boys in my chemistry lab stayed after school to create their own rocket club. They met in an open field to launch homemade rockets. In class, they were eager to share what they had built and launched, but they never invited me to join them, which was just as well. After school, I had to take care of my brothers and help them with homework, also making time for my own studies.

I tried to keep up with the washing, ironing, and cooking, and I babysat the neighbors' kids for spending money, but it seemed that I always failed at something. Daddy worked hard and got a better job building new homes, and when he was able, he helped me cook, frequently telling all four of his children that we did a fine job of keeping our family together.

One morning as I rushed to get us all ready for school and work, Daddy complained, "This coffee tastes like axle grease."

Not surprised, I confessed, "To save money and time, I've been using the same coffee grounds, adding only about a teaspoon of fresh coffee to the pot each day."

"Hmm. You shouldn't do that, June Bug. I'll be making the coffee from now on." For once, I appreciated his reprimand because it helped me cross a chore off my crowded list.

ELATION AND DEVASTATION

Every day I wrote a letter to Mother, telling her about our lives, expressing how we missed her, and hoping she would write back and say she was coming home. Late in the spring, after months away from us, she was allowed to return. With cheerful smiles and big hugs, we welcomed her and celebrated the night with a "Welcome Home" cake that I baked. She wanted to hear about our holidays and school and friends. Three-year old Lee climbed into her lap and snuggled in her arms. "Me too!" we all said, wanting to love on her and needing her to love on us.

When the time seemed right, I told my parents I wanted to find a job so I could work after school and earn money for school clothes. They agreed, so I went into the city and searched for a part-time job. Walking the streets, stepping into shops, asking about a job, and learning they weren't hiring was discouraging. My feet ached, and so did my confidence, until I saw a "Help Wanted" sign in a shop window. Feigning confidence, I put on a cheery smile and walked through the massive doors of the S. H. Kress & Company, a "five and dime" retail department store. Using my best grown-up voice, I asked for an application.

A few minutes later, the manager put on wire spectacles, wrinkled his nose, and read what I had filled out. With a huff, he said, "So this is your first job other than babysitting and general housecleaning?"

"Yes, sir!" I answered, standing at attention, eagerly waiting for his response."

"All right, young lady. We'll hire you on a trial basis to work after school and on Saturdays. If we like you, there may be a full-time position for you during the summer months."

When school let out for the summer, I was rewarded for my dependability with extended hours. Eager to cash my first big check and put it into a savings account, I mentally added up how much I could earn with my weekly salary. I was delighted to discover that I would have enough to buy school clothes in the fall for senior year and still have money left to save for college.

That summer in 1958, just before my sixteenth birthday, I followed the news about Congress and the Eisenhower administration's plan to set up the National Aeronautics and Space Administration (NASA) as an agency for the government of the United States, responsible for the nation's civilian space program. All seemed well and good on my little part of the planet.

However, joy soon turned to desperation. Late one afternoon while Daddy was still at work and I helped Mother with dinner, she said, "I've worked it out with our neighbor to take me and your brothers in her car and move us back to San Antonio. You can come too if you want."

The earth fell away under my feet as my legs went limp and my eyes filled with tears. "Why?" I asked. "We're happy, and if we leave Daddy to go with you, it will be the same as before. We'll be homeless and scared. That's no way to live."

She stepped from the sink with her wet hands and smacked me hard across my face. "We don't need you or your smart mouth to go with us."

Undeterred, I pressed, "But what will happen to my brothers when you have to go to the hospital again?"

Mother's glare shot through me with daggers, and her arms stretched out and grabbed fists full of hair as she tried to strike me again. With tears flowing, I ran out the front door and headed to Kathie's house for refuge and sanity. She and her mother consoled me and encouraged me to stay with them until Daddy came home from work. Kathie and I huddled together in her bedroom, promising to be best friends forever no matter what happened. We even made a pact to name our daughters after each other.

The following morning, I left early to catch the bus for work. I left home without seeing Daddy. That day, I stood behind the counter where we sold hair products like combs, barrettes, brushes, and sewing items. After making my first sale, I plunked the few coins into the cash register and looked up to see Daddy standing there.

I hadn't seen Daddy with tears in his eyes since the last day he saw his mother before she died. As he began to speak, the tears spilled out and rolled down his suntanned cheeks. "Please don't go with your mother, Junie," he begged. "You are old enough to stay. You can finish your senior year in school. Soon we'll have enough money to buy a home and pretty furniture for your own bedroom."

When my choice came down to whose heart I was going to break— Daddy's or Mother's—I deeply felt the conflict between what I wanted to do and what I knew I had to do. I had to go with Mother so I could care for my brothers.

BACK TO SAN ANTONIO

Though heartbroken, Daddy understood. He asked me to send him our new address so he could mail money as he earned it. As he walked to the door, I ran out from behind the counter to hug him good-bye.

That afternoon when I returned home, the neighbor, Jewel, had filled her car with what she needed for the trip and what my mother wanted to take to San Antonio. Jewel said she would wait for me to get my things. Inside the house, I hurriedly packed my little Bible, some clothes and toiletries, and Norman Vincent Peale's *The Power of Positive Thinking*, my favorite book I had carried with me since I was nine years old. Within minutes, we drove out of West Palm Beach toward the unknown. I left my yesterdays behind and carried with me the memories of our time in Florida and the sorrow of leaving Daddy alone, with only his tears and the dreams for his family left unfulfilled.

When we stopped for the night at a roadside motel, I walked beside my mother toward the door of our room, put my arm into hers, and whispered, "I don't understand, but I'm sorry for what I said last night."

She said nothing as she unlocked the hotel door, pushed it open, and dumped an armload of things on the bed. She took her pack of Camels from her purse, then walked to the sliding glass door. Standing there, she pushed the door wide open and looked out into the distance. Then she lit a cigarette, breathed in deeply, and replied in an exhale of smoke, "I will be happier in San Antonio. I'll get a job. We'll make it work out for us. You'll see."

Later, my brothers turned on the television and piled in the bed to watch a superhero program. Mother and Jewel sat at the table with their cigarettes and soft drinks, and I asked if I could step outside.

The night air felt cool and fresh as I walked around a garden of shrubs and fragrant flowers surrounding a small swimming pool. I sat at the edge of the pool and dangled my feet in the cool water. On the surface, I saw the reflection of a star, or maybe it was Venus. With tears blurring my eyes, I couldn't be sure. Quietly, I recited, "First star I see tonight, I wish I may, I wish I might . . . someday . . . ." My voice trailed off as I stopped to think. Then I prayed in desperation, begging God to help me escape the chaos of my family life and guide us as we tried to find our way.

Chapter 5

STAR LIGHT, STAR BRIGHT

A TEXAS WELCOME

Late in the summer of 1958, with the car packed to the ceiling and us squeezed in, we drove into San Antonio with nowhere to go. Mother asked Jewel to drop us off anywhere. Jewel insisted on taking us to a friend or family member. At a telephone booth, Mother looked up the phone number of Gordon Perry, her brother, and then called him for his address. Jewel dropped us off on Uncle Gordon's doorstep, each child holding a bag of simple treasures, and Mother carrying her box of items with the sage and rose bedspread folded on top.

It was hot and sultry every day of the month of August. Uncle Gordon, Aunt Gladys, and two cousins welcomed us into their home with kindness and love. Our extra five people made their two-bedroom home bulge with activity. Even so, I felt loved with family surrounding us. They even provided a small table radio for our bedroom, so I could listen to my favorite radio stations.

Mother and I shared a guest bed, while my brothers made beds from old quilts on the floor each night. Both Gordon and Gladys worked during the day, so we tried to help around the house and prepare meals for them when they returned home.

Before long, Mother grew restless and wanted to move to the other side of town to live with our grandmother in her one-bedroom house. She was sure she could find work there. Uncle Gordon helped us pack and move to Grandma's. She welcomed all five of us with outstretched arms and cheers of delight. She spread out a feast of food. Even though Papa had died several years before, Grandma still had a vegetable garden and chickens. Her years of experience raising her own nine children made our time with her extra special.

EDUCATION AND EMPLOYMENT

Within a few days, my brothers and I registered for school. The counselor looked at my Florida report card, noting my courses and grades. She suggested that I continue with the course work for college-bound students. It sounded wonderful, but I hesitated, suddenly realizing I might have to put my college dreams on hold. Mother had advised me to take practical courses like typing and shorthand so I could get a job as a secretary after graduation.

"Is that what *you* want to do?" the counselor asked.

"I don't have a choice," I responded. "I need to help my mother and my three brothers."

The counselor wrote in five subjects for me to take, including Texas history. I explained that I'd already studied Alabama and Florida history, but she laughed and said the class was a graduation requirement.

I didn't think it was funny. In fact, I was upset when I looked at my subjects list. Where was the *Sputnik*-inspired curriculum introduced to us in Florida? With all the requirements for learning to be a secretary, my schedule left no time for math and science courses. Aggravated, I scratched through one of the secretarial courses, left typing as a subject, and asked which advanced math or science class I could take. The counselor added a combination class of advanced algebra, trigonometry, and calculus and suggested that I help create a science club after school to do some of the physics activities I enjoyed.

As I walked the halls of Harlandale High School, I thought it seemed strangely unfamiliar—nothing like either the small, rural Alabama high school or the giant campus of the city high school in Florida. The kids and teachers were great, but there was great emphasis on football and cheerleading activities.

Studies were uncomplicated, so I decided to look for a job. Since I rode the bus, Mother offered to drive me home after work in our 1947 Chevy, which I just knew was the oldest car on the road. It used a gallon of oil to every gallon of gas, and fumes from the tail pipe spit out oil like rain.

A post office stood near the school, so I stopped in to ask for a job application. The two men looked at me with surprise, turned to each other laughing, and said, "The post office does not hire women."

I was insulted, shocked, and indignant. I responded, "Did you know our country has entered the space age? This is the twentieth century! Why can't a woman stand behind a counter and sell stamps and weigh packages?"

The two clerks laughed uneasily and shook their heads. Dejected, I walked home nearly seven miles, but on the way I saw a help wanted sign at a day care adjacent to a motel office. I hoped they had a job opening that would allow me to work with children, but the job was for cleaning bathrooms and washing towels and sheets. I took the job anyway, knowing that if Mother didn't find work, we'd be struggling to survive.

A NEW CHURCH AND NEW FRIENDS

Grandma insisted that we all attend church with her each Sunday at Mayfield Baptist Church. It was wonderful to be part of a church family again and meet new friends. I already had a dear school friend named Ellen, and I met another Ellen at church. Each Sunday during Sunday school, she whispered stories about her date with Jake, a boy in the Air Force.

In October, just before Halloween, our church organized a hayride for the teenagers. We were encouraged to invite friends to join us, so Ellen asked her friend in the Air Force to bring one of his friends to meet me. The night of the hayride and cookout on the river, Ellen and I waited for her friends at the church. Driving up in a green '51 Chevy, the two boys pulled into a parking space, stepped out talking seriously about something, and then walked up and courteously introduced themselves. I think Jake, Ellen's friend, said hello to me, but I only saw the tall boy with dark hair and deep blue eyes.

"Hi, I'm Jake's friend," he said, grinning shyly. "My name is Dick Scobee. Thanks for inviting me to your hayride."

Noticing his green football letterman's sweater, I asked, "What's the big 'A' stand for on your sweater?" I wondered if he was a former Auburn student.

He explained that he had attended Auburn High School in Washington State, playing football for them until he broke his wrist.

After that brief conversation, the youth leaders motioned us to climb aboard the wagon hooked to the back of a truck. Dick boosted me onto the hay, and then we made our way to the front to sit with our backs toward the truck cab. All the way to the picnic area, we talked about our families. His younger brother, Jim, was the real athlete in the family. His father was a railroad engineer in Auburn, and their family had lived in only two different houses and attended the same school all their lives. I was incredulous.

Dick was nineteen years old, and I hesitantly told him I was only sixteen, though I was a high school senior. I explained our frequent moves and that I'd been held back a grade but moved up at least two. Then I mentioned that my parents were divorced and that I lived with Mother and my younger brothers at my grandmother's house. (Mother and Daddy divorced and remarried several times over the course of their relationship.) Not wanting to get further into my family problems, I awkwardly changed the subject. "Since you're in the Air Force, what do you think about the new space agency called NASA?"

He'd heard of it, but he said the rockets were too fast for him. Dick liked jets and propeller-driven planes. He dreamed of attending the Air Force Academy but was told he couldn't apply because he didn't personally know a congressman who could recommend him.

"So here I am," he said, "just an airplane mechanic at Kelly Air Force Base. That's the closest I could get to an airplane."

I told I had lived near Huntsville where they built the rockets and in Florida near the Cape where they launched them, and explained that I had studied aerodynamics and airplanes when I was in the fifth grade. At that point, I wasn't ready to explain my year at the foster home and the "A" encyclopedia that I "borrowed" and kept hidden under my bed.

When we arrived at the picnic area, the wagon filled with bales of hay and bouncing teenagers parked at a grassy field surrounded by trees. Dick Scobee jumped off the wagon bed first, and then reached up his hand to help me down. Surprised, I blushed at the attention.

STARS IN OUR EYES

As the others scattered to begin building a fire and setting up camp, Dick and I were more interested in getting to know each other. We walked toward the small river, which babbled over stones. The setting sun fell behind the horizon as the sky darkened to night and stars unfurled overhead. We both looked up and noticed the stardust of the Milky Way.

"All those stars in our galaxy make me feel very insignificant," Dick commented.

I gestured toward the Big Dipper and named other constellations, then asked, "What's the chance that we'll see a satellite fly over?"

He made a beeping sound like orbiting satellites, laughed, and naturally took my hand to lead me across the grassy field toward a giant oak tree. We leaned against the tree and talked about our dreams and wishes and hopes. I was giddy. I told him I didn't ever want to grow up.

"I want to fly like Wendy in Peter Pan to Never Land," I said.

"I want to fly too, but in real airplanes," he said seriously. "Would you fly with me? If I learn to fly and can get an airplane?"

"Sure thing!" I responded gleefully. Then I shared my desire to go to college and learn to be a teacher who inspires children and makes learning fun. Suddenly, he turned to face me, brushed my hair aside, and lifted my chin. He gently kissed my forehead, then each cheek, and ever so quickly brushed his lips against mine. I thought the heat upon our faces could ignite the bonfire wood stacked behind us.

Rather shakily, I suggested that we rejoin the group.

All too soon, it was time to smother the fire and watch the hot embers cool before we climbed back into the wagon and returned to the church. On the drive back, I grew cool wearing only a light sweater, and when I began to shiver, Dick draped his bulky letterman's sweater across both of us, but only after he pulled it above both our heads and grabbed another quick kiss.

Back at the church, the four of us walked to Dick's car, which he and his friend had to push out of the parking space since it couldn't go in reverse. Before he drove away, he asked for my grandmother's phone number and wanted to know if I would go with him to a drive-in theater on Saturday night. Thinking quickly, I realized I was both afraid to be alone with him and anxious to see him again, so I answered, "Yes, I'd love to go with you if you'll come to church with me on Sunday." Then I added nervously that my younger brothers would love to go the movies.

Hesitating only briefly, he said, "You bet! Yes, on both accounts!"

I stood dazzled and euphoric as he waved good-bye, then turned to Ellen who was grinning like a Cheshire cat.

I was speechless, and on the ride home I wondered if I had met Dick by some quirk of luck or by God's goodness. Already, I felt that God had blessed me far beyond any of my childhood dreams to find a good friend.

After talking on the phone the next evening to settle details, I waited eagerly for Saturday. When the day finally arrived, Lonnie and Johnny teased me unmercifully about the date and how they would act as our chaperones, but they admitted that they were excited about the rare opportunity.

I found it awkward to introduce Dick to my mother. Concerned that he was driving three of her children, she grilled him on safety and insisted knowing more about the movie, an Andy Griffith comedy called *No Time for Sergeants*. When Mother learned the subject matter, her eyes glazed over as she fearfully began telling stories about how her brothers survived the war in Europe and later Korea War. She tossed out words of anger for the Communists and Nazis. Dick responded respectfully and seemed curiously interested, but as she talked, we all backed cautiously toward the door, hoping she wouldn't change her mind about letting us go.

A GROWING FRIENDSHIP

During the cool November days, Dick Scobee and I saw each other regularly. Like tourists, we visited the historic sites of San Antonio Missions and the Alamo, talked about why he joined the Air Force and what his high school friends were like, and discussed his family. He missed his mother's fresh baked cakes and berry pies, but he didn't miss the rain that fell from fall through spring on the west coast of Washington in the valley where he grew up in Auburn, near Seattle.

Dick's first job was working in the fields, picking beans for money to buy new school clothes. Later he worked as a bag boy in a grocery store. After high school, he took a job at Boeing Air Field until he could enlist in the Air Force. With his high score on the entrance exam, he was asked to go into the field of Intelligence, but he requested to be assigned closer to the flight line and airplanes.

Whenever questions arose about my parents, I hesitated to tell the whole story, fearing that it would influence his opinion of me. He often invited my brothers and mother on our excursions. One of our favorite sites to visit was the Japanese Tea Garden at Breckenridge Park on the north side of town. It was like a winter playground with waterfalls, plants, and koi fish ponds with floating water lilies. Dick and I held hands as we walked the fairy-tale paths across the unique stone bridges. My youngest brother liked the miniature railway, and Mother loved the Chinese Sunken Gardens.

On one of our last outings before Dick left San Antonio for a few weeks of military leave, we drove to the airfield and parked near the end of the runway. We shared what we knew about airplanes, talking about the wings and the forces of lift and drag and varying designs of fuselage with lighter materials. I told him what I knew about Chuck Yeager and his famous flight that broke the sound barrier. Our discussion led to college.

Dick put his arm around me and confessed, "I'm probably not smart enough to go to college, and I sure can't afford what the tuition costs."

I offered what I knew about the boys in my school who had already been accepted to universities. "You are so much smarter than some of the boys I go to school with, and you can start out at a community college and work at the same time."

He grinned. "*You* are the smart one. I saw your report card with all A's on every subject line. If I can go to college, then we'll both go together."

Dick sealed the promise with a kiss that stirred unfamiliar emotions and caused me to panic. I asked to return home.

That night on my grandmother's front porch, we hugged and kissed good-bye, since we wouldn't see each other until after Christmas. As he turned to leave, he said, "Tell your brother thanks for the picture he gave me. It's of you. I put it in my wallet, and I'm taking it with me."

During the holidays, Mother became disgruntled with Grandma and began to speak of her the way she'd often spoken of Daddy. She announced that we would move out and rent a house she found on Commercial Avenue, about five blocks away.

Just after Christmas, we moved into the simple white frame house with four rooms, a bathroom, a carport, and a storage shed. The home was fine, but I missed Grandma and visited her often to chat and to use her Singer pedal sewing machine. We were concerned about Mother's mental state, but we enjoyed sharing stories and memories with each other.

THE REALITY OF LOVE

During one of our conversations about a particularly trying incident with one of my cousins, Grandma reached across the table and placed her hand on mine. "Those are great memories," she said, "some sweet and some that hurt. Life's like that. Life's about learning lessons, and if we don't learn them, we live in regret." As she got up to place our empty teacups in the sink, she added slyly, "That boy Dick Scobee likes you a lot. There aren't many out there that smart and nice, you know."

"We might be getting too friendly," I admitted. "Maybe I shouldn't see him anymore. I should make friends with other people, and besides, someday I'm going to have to tell him about Mother's mental illness, and he'll quit seeing me anyway."

Surprised, Grandma protested, "He's a nice young man, and very good to you and your brothers. If you tell him the truth, you may find he's understanding about your situation. Your friendship could grow even closer."

I figured she was probably right, but growing closer to Dick scared me too. As I left Grandma's, I told her that if Dick called, she should tell him I had moved and was seeing other people.

Saddened by my decision, I began walking home, pulling my coat collar high around my neck and face as the cold wind seemed to chill the depths of my soul. I felt heavy of heart, and it reminded me of the way I felt when we had left Daddy in Florida the summer before. "It can't be the same," I argued with myself. "That was *love*. This is infatuation." But then I told myself that ending a mere infatuation wouldn't hurt so much. Clearly, there was more to Dick and me than that.

Chapter 6

HEAVEN AND EARTH

BLOSSOMING LOVE

The last semester of my senior year flew by after the Christmas holidays. I frequently met to study with my sweet school friend named Ellen. One day, she gathered all the girls to share "fantastic news." In a grand gesture, she raised her left hand to our faces, displaying a diamond on her ring finger. We were all ecstatic, cheering and asking more questions than she could answer.

"It will be a June wedding!" she finally exclaimed. "You are all invited." We all hugged, everyone genuinely happy for her.

In 1959, Valentine's Day fell on a Saturday. On Friday the day before, I sat in my typing class and struggled to find the words for a friendly note of apology to Dick Scobee. He deserved an explanation for my sudden change of heart, but I couldn't decide how to put it, so I typed out a brief "Hello, hope you are doing well." After putting the letter in the city mailbox, I knew it was too late to alter my impulsive act. I shrugged and turned to go back to my afternoon classes.

Saturday morning, my family greeted each other with homemade valentines. I bought my brothers small heart candies and a big piece of chocolate that they wanted to eat before breakfast. When I saw the mail carrier outside, I stepped out to reach for the letters. It was mostly utility bills, but I also found a large envelope addressed to me with no return address. I opened it to see a heart-shaped valentine that read, "To someone too nice to forget!" There was no signature. The only person I knew who was thoughtful enough to do something like that was Dick Scobee—and he was too proud to sign his name.

My mind went into a tailspin, wondering how it was possible. I imagined our notes passing each other in the mail and considered the thought that we might be destined for each other. With my heart pounding, I twirled in delight, clutching the valentine to my chest.

On Sunday, I went to church with my grandmother. Just before the service began, someone in a blue tweed sports coat and tie asked if he could sit next to us. It was Dick Scobee. Grandma said, "Of course, please join us." Turning to me, she grinned and murmured, "That smile on your face is bright enough to blind us."

For long moments, Dick and I simply gazed at each other, blushing and smiling. When we stood together with our hymnal books, he asked, "Did you get a valentine card in the mail?" I nodded and asked if he got a typewritten note. He smiled in reply.

The sermon was based on 1 Corinthians 13, the letter about love that Paul wrote hundreds of years ago. The pastor emphasized, "Love has its risky side, its happy side, and also its painful side. There are no guarantees."

As we left the church, Dick asked Grandma if he could drive me home. When she nodded approval, he rushed me to the parking lot to see his new car, a blue and white '55 Ford Fairlane. With great fanfare, he opened the passenger door, guided me into the car, and said, "I have another surprise! This car comes with all kinds of special amenities, including a transmission with reverse." We laughed at our joy of being together. It felt completely natural, even after many weeks apart.

Spring came, warming the weather along with our hearts. Our conversations were natural. Our values were similar. We encouraged each other's hopes and dreams. Dick came to school functions with me, picked me up for lunch in his car, and took me swimming at the lake and walking in the park.

On one of our more memorable dates, Dick asked me to go to dinner and a movie with him. As we left the restaurant after dinner, a boy whistled at me. Dick's face turned red as his temper flared. He walked over to the guy and threatened, "Hey, the girl's a lady, and the lady's mine."

His temper mortified me, but his insistence that I was his girl petrified me. I didn't know what to think or say. I was embarrassed and surprised but also comforted by his protective defense of my honor.

The movie we saw wasn't as memorable as it was to see the Majestic Theater in the heart of downtown on Houston Street near the River Walk.

The interior was a fantasy villa with walls and towers built in the Mediterranean style. Walls were painted with grapevines, but grandest of all was the vaulted "sky" with drifting clouds passing overhead that actually came to life with stars twinkling overhead when the theater lights dimmed. The experience was heavenly.

At school the following day, I walked in a daze, wondering how I could possibly be any happier. I'd never felt so loved and cared for in my life. I counted the seconds until I could go out with Dick Scobee again.

A FIERY DISASTER

As the school bus slowed to a stop near the front of our house, the other students shouted that the back of our house was on fire. I jumped off the bus and ran to the backyard to see what was burning. When I turned the corner of the house, I saw my mother tossing items into a barrel blazing with fire.

"What's in the barrel?" I asked frantically. "What are you burning?"

She backed away, wiped the sweat from her smoke-streaked face with her forearm, and glared at me. "Don't you hear them?" she whispered.

"What? Hear what?" I shouted, shaking my head.

As smoke and ash circled around our heads, she asked again whether I heard the voices telling her to burn my clothes and our photos. Pointing to the blazing barrel and the box that lay on the ground next to it, she explained, "The voices are foreign. Can't you hear the *oriental* voices telling me to destroy anything that would identify us with our past?"

Peering into the cardboard box, I cried out in horrified shock, "Not our family pictures!" I reached into the box, but none remained except for three photos stuck in the folds at the bottom.

Mother screamed, "Communism is on the rise! Out in Hollywood, the movie industry is a hotbed for Communist spies and sympathizers. They'll be looking for us. They'll find out who we are if they can identify us by our clothes in the photos."

I had no words to reason with her. I knew she had stopped taking her medicine and asked her to come inside to rest. Once my brothers were settled in for the afternoon, I ran to Grandma's house to tell her and call Uncle Gordon, whom I hoped would know what to do about Mother.

My uncle gave his familiar sigh of hopelessness, but he calmed me, saying he would contact her doctor and then come to our house on the weekend to take her to the clinic for her medicine and therapy.

Back at home, my brothers and I whispered about Uncle Gordon coming to help. Trying to maintain a sense of normality, we had dinner and talked about schoolwork as if nothing had happened. The following morning, we went to school and Mother went to work—just like every day—though my brothers and I had to explain the fire to the other kids on the bus. We joked about burning leaves and laughed it off as nothing.

That afternoon, we faced more drama as we stepped off the school bus. With faces red from embarrassment, we ran up the driveway to see all our belongings and household items tossed outside under the carport, spilling over into the grassy yard. A lampshade stood on top of a pile of shoes. Bed linens were tied together with clothes and towels. A toaster sat on the doorstep on top of a few books and my little Bible. Toys and kitchen supplies were scattered as though they were thrown out the door in anger. In wonder and humiliation, we stood before the locked front door and read the notice stapled to the wooden frame. It said we were evicted from our home.

I explained to the boys that Mother probably hadn't paid our rent, so the landlord threw us out of our house and locked the doors.

We were all angry and resentful, and then I suggested that they wait at the house with our things while I went for help.

BACK TO GRANDMA'S

By Saturday, boxes of our things filled a room at Grandma's house, and Uncle Gordon had taken Mother to the San Antonio State Hospital. My mind was spinning with confusion, my stomach ached, and my eyes burned with unshed tears. It was hard to believe this was the end of the same week that began so beautifully. Completely focused on mere survival, I was surprised to see Dick Scobee drive up at Grandma's. He raced to the front doorstep where I stood waiting.

Reaching for my hand with confusion on his face, he asked, "What happened? I went to your house and you were gone."

Reluctant to go into details, I braced myself to tell him what had happened, but my brothers beat me to it, shouting out all our tribulations before I could stop them.

Before I could explain, Dick took me in his arms, caressing my hair and reprimanding me. "Why didn't you call me to help? That's why I gave you my phone number in the squadron and told you to call any time."

I looked into his piercing blue eyes, which were filled with concern. At that moment, I knew he really loved me. I let tears flow down my cheeks, soaking my face and his shoulder. When he was satisfied that everyone was going to be okay, he asked me to go to dinner with him. We were both quiet in the car. At the drive-in, we ordered burgers and Cokes curbside from the girl on skates, then decided to change our order to chocolate milkshakes. At first, it was difficult to swallow, so I just sipped.

Finally, Dick asked the question I had dreaded for months. "What is your mother's illness?"

TELLING THE TRUTH

I gathered courage, breathed a heavy sigh, and began to tell Dick about Mother's schizophrenia and our many family problems.

He listened carefully as I explained the pain our family had endured over the years, which had ebbed and flowed along with the relapses of Mother's illness. I told him about the truant officers who came to our house when I was eleven and learned that Mother kept us out of school to protect us from Communist indoctrination. I told him about her many hospitalizations, medications, and electric shock treatments, which at the time were thought to help the mentally ill. He listened as I described my mother's brilliance and beauty that quickly faded during one of her terrible episodes. He learned about the difficult days when Mother was away and I had to take care of the house and my brothers. I told him about the rejection we felt from other kids and even adults who didn't understand our situation. I mentioned how anxious Mother got over any news of war and told him about moving often, leaving our beloved toys, being homeless, scrounging for meals, and living without Daddy.

Cautiously, Dick asked, "Are you angry or bitter about all those moves and problems with your parents?"

"What a great friend you are to care and to ask!" I exclaimed, not answering his question.

"I've heard it's best if you don't let the sun go down on your anger," he said. "The courage to forgive and move on is liberating, though it's easier said than done. Did you ever read any Ernest Hemingway? He wrote in *A Farewell to Arms*, 'The world breaks everyone, and afterward, some are strong in the broken places.' But you're still so feisty. I'd guess that your spirit's never been broken."

"Well," I responded, "let's save that story for another night. I've talked way too much."

On our ride back to my Grandma's house, we sang along to our favorites on the radio—Pat Boone's "April Love," and Frankie Avalon's "Venus."

Snuggled next to him, I looked out the car window to the stars that seemed to shine brighter than ever. "Venus was the 'first star' I wished on each night before I found out that it was really a planet," I confessed.

"I think it's too late to see Venus tonight," Dick said, "but we could stay out all night and watch for it first thing in the early morning before daylight. Isn't Venus named after the Roman goddess of beauty and love?"

"Good try, Scobee, but you'd better get me home before the clock strikes midnight and this fine blue and white coach turns into a pumpkin."

He looked at me and laughed.

Chapter 7

SUN-SPLIT CLOUDS

THE *MERCURY* SEVEN

In April 1959, the selection of the *Mercury* Seven astronauts headlined newscasts and other media. The *Mercury* program was the first human spaceflight project of the United States, and its goal was to put a human in orbit around the Earth. The photographs for newspapers and the cover of *Life Magazine* showed seven men clad in space suits and helmets that resembled something out of science-fiction books, but they were real. Stories about them, their hometowns and childhoods, and mostly their plans to fly in space showed that the United States was serious about manned space flight. I told Dick that some of the boys at school asked if my Auburn friend in the Air Force knew anything about the Mercury project and astronaut qualifications.

Dick laughed. "One thing I know for sure is that I'll never qualify," he said. He had followed the project announcements closely, noting that the first administrator of NASA, Keith Glennan, needed astronauts who were younger than forty years, were shorter than 5 feet and 11 inches, and were jet pilots who had graduated from Test Pilot School.

"I'll be happy if I ever get a chance to fly airplanes," Dick told me "Those guys are brilliant college graduates and test pilots, and besides, you forget I'm 6 feet, 2 inches tall."

A LIFE-CHANGING SURPRISE

One day in May, with the school year drawing to a close and the temperature rising, Dick called my grandmother's house to say he was picking me up at school and taking me out for lunch. He worked as a mechanic on C-124 Globemaster cargo propeller planes, often traveling overseas on them to work with the crew and earn extra money. He was in town for the day and wanted to see me.

When the lunch bell rang, I tossed my books into my locker and hurried toward the front door. Out the big swinging doors, I ran down the sidewalk to the school parking lot. The air felt misty, and dark clouds were forming overhead.

Dick's blue and white Ford was parked on the curb, where he waited for me with a hearty grin. "I have a surprise for you!" he said.

Soon he pulled into our favorite small café. I saw him reach under the car seat and put something in his pocket but thought little of it. Inside, he walked me away from everyone toward a quiet area at the front of the restaurant. We sat at a small table near a window, facing each other and grinning joyously.

While Dick gave the waitress our order, I glanced to the sky to see that a ray of sunshine poked through the clouds and shined right on us. Mesmerized, I said, "Look, it's like a golden beam of light shining straight through those dark clouds."

After the waitress delivered our hamburgers and Cokes, I noticed that Dick was red-faced and anxious. From his pocket, he retrieved a small box and opened it for me to see. I could not believe my eyes. It was a beautiful, delicate, solitaire diamond engagement ring.

"I know we're young," he began, "but I love you and I want to live my life with you forever. I'd be good to you. We can get married this summer after you graduate, or we can marry after college." He gave me another broad smile. "Well, will you? Will you marry me?"

I was stunned and bursting with joy at the same time. My voice barely audible, I protested, "There's so much going on with my family. I need to help them before I can think of myself." As we looked into each other's eyes, the world seemed to disappear. "Are you sure?" I asked. "Do you really want to marry into this crazy family?"

"I love you, June, and everything that goes with you," he assured me. "I promise by the moon and the stars that I'll always love you. We can make it work. Please say you'll be my wife."

"Yes!" I finally exclaimed. "I've loved you since that first kiss the night of the hayride." The words bubbled out of my mouth as tears trickled from my eyes. Reaching for a paper napkin, I wiped my cheeks and blew my nose, then laughed at my unromantic gestures. "You truly are a dear and wonderful man. Yes! I love you too!"

He placed the ring on my finger and proclaimed, "We are officially engaged to be married!" Then he reached across the table and caressed my face, joking, "I want you to wave that ring around at school where all the boys can see that you are taken, that you are mine."

As we kissed, we grew aware of everyone looking at us. Turning our faces away from the other customers, we realized we had eaten none of our lunches, and it was time for me to return to school. We quickly gobbled down a couple of bites, paid the bill, and rushed to the car. Dick had to hurry back to the base to pack and prepare for another Air Force trip overseas. Even in our hurry, nothing could dampen our spirits.

ON CLOUD NINE

Back at school, I walked to class as if on a cloud, with my mind spinning and feeling dizzy. I stumbled from one class to the next until I met my friends after school.

My friends were unbelieving at first, but as they studied my ring, they broke into cheers of shared happiness. They told me how expensive a wedding could be, how much I had to plan and prepare, but my dear friend Ellen mentioned that if we married after her June wedding, she would save everything possible for me, including her wedding dress.

Dick had already spoken with Grandma, and together we wrote a letter to Daddy. Dick had to leave for a military flight overseas, so I shared the good news with my brothers and uncles. Grandma was happiest of all for us. When I told Uncle Gordon I wanted to get married at the church, he suggested that we elope to save money. I insisted that our wedding needed to be in the church and blessed by God.

When Dick returned from his travels, we made plans. The local minister reserved a Saturday in July for us. Since I was underage, we needed parental permission to get a marriage license from the courthouse. When Daddy received our letter, he had immediately called Grandma to say he was driving out to San Antonio to meet this boy, and he wasn't waiting for the wedding. He wanted to come for my high school graduation. "I won't miss seeing with my very own eyes my June Bug's graduation," he said, "and what's this about graduating near the top of your class?"

GRADUATION DAY

Dick and I got an application for our marriage license from the courthouse and were advised to use a typewriter to fill in our information. The clerk also reminded us we needed a signature to grant permission for me to wed. With little time to spare before Dick had to return to work, we ducked into a typewriter store where a kind salesman lent us one and hollered out congratulations as we left with the completed application.

With that item crossed off our to-do lists, we faced graduation day. Daddy, who'd arrived earlier that day, wanted to drive Dick and me into the city for the graduation ceremony. Dick and I both noticed that he was breathing heavily and driving with difficulty. The ceremony was long and involved, but both Daddy and Dick told me how proud they felt when I walked across the stage to receive my diploma and award for being one of the top graduates. On our way home afterward, Daddy announced that I was the first in a long line of family members who would go on to college. He was excited for my future.

Later that night, I mentioned to Dick my concern for Daddy's health. He affirmed my concern, but we figured he was probably just tired from the trip. The next day, Daddy talked to Grandma about our pending marriage and asked her what she thought. Fortunately, she admired Dick and told Daddy, "If Dick Scobee is anything, he is a serious young man. I believe he truly loves your daughter, and he's been good to our entire family."

At the notary public's office, Daddy signed the marriage certificate, giving permission for me to marry. Then, blowing his nose with his big white handkerchief, he said, "Please come see me in Florida for a visit. I want to show Junie how I fixed up her bedroom." As he turned to leave, he added, "I'm going to stop by to visit your mother in the hospital. I'll ask her if she wants to bring the boys and go back to Florida with me." I hoped they'd have a happy reunion.

THE CHURCH WEDDING

On our wedding day, we got a break from the hot and humid weather as a light breeze freshened the air. Family and friends gathered at the reception hall to help us organize our ceremony at Mayfield Park Baptist Church.

After he gave us permission to marry, Daddy, who knew he was quite ill, had asked dear Uncle Gordon to walk me down the aisle. I was happy to have him, as he'd been like a father to my brothers and me.

"Your dad would be proud of how pretty and happy you look," Uncle Gordon said, smiling at me in my creamy white wedding dress with a suggestion of pink in the silk underskirt, like the pink on the inside of a seashell.

As the pianist played the wedding march, Uncle Gordon slipped my arm into his, patted my hand, and asked, "Are you ready to walk up the aisle to meet Dick Scobee waiting for you at the altar?"

"Oh, yes," I insisted, stepping out softly.

The pastor read the vows proclaiming the importance of marriage, and then elaborated, "For a marriage to be successful, you must develop a team spirit, each cheering for the other in life, work, and play. On your journey together, you will have bumps along the way in the road, but how you handle those bumps determines whether your love for one another will grow stronger or wither and fade away with each passing day."

After the ceremony, we gathered in an assembly area adjacent to the church where my grandmother waited to show everyone the tiered wedding cake she had baked and decorated. Mother was there, too, smiling happily for us, while my brothers grinned as though they had done something mischievous. And indeed they had—Dick's car was smeared with shoe polish wedding wishes, complete with cans and shoes and unmentionables tied to the back bumper.

After the wedding, we drove across town to our one-room garage apartment that boasted a tiny kitchen to one side and an even smaller bathroom on the other side. Though small, it was nicely furnished and a welcome place to begin life as a couple. I felt safe and could not have been happier if it had been a mansion, especially when Dick carried me over the threshold and announced, "Welcome home, Mrs. Scobee!"

The next day, we had a long drive from San Antonio to his hometown near Seattle, Washington. Instead of his parents paying to fly to Texas for the ceremony, they chose to pay for us to come to them for a family reception in Auburn. I looked forward to meeting them.

JOURNEY TO AUBURN, WASHINGTON

The following morning, we left before sunrise to drive northwest. Though I had often traveled the highways in the Southeast, this was my first time to travel north. I enjoyed seeing new sights and experiencing an adventure beyond my wildest dreams. We slept in hotels and ate at restaurants on the way. In the car, we sang our favorite songs and talked about our future, in particular what I might expect when we arrived in Auburn.

After a few days of travel, we drove into the small railroad town of Auburn and down Main Street. Dick could not contain his enthusiasm as he told me about every landmark and memory. We drove past his high school, the Safeway store where he had worked, the movie theater and his favorite hamburger drive-in, the Rainbow Cafe. We turned the corner onto Fourth Street and drove up to the front of the Scobee home, a modest, two-story wood frame house on a slight hill surrounded by luscious green grass and vibrant flowers. We were greeted at the door with the welcome and warmth of family that I'd only dreamed of having. I adored his parents, and they showed me considerable kindness and loving words of acceptance.

Dick took me to his bedroom to show me his collection of airplane models hanging from the ceiling, his comic books and airplane magazines, and a couple of football trophies. We rested, bathed, and dressed for an event beyond words. His family, extended family, and friends filled the rooms of the house, spilling out to the garden and back yard, where a giant cookout and reception waited for us. As we stepped out the door, they cheered to see us blushing with embarrassment. First Dick's dad, then his uncles insisted on hearing me talk. "But why?" I wanted to know.

"So we can hear a real southern accent," they chorused one after another.

Looking to Dick for a rescue, I found that he offered no help. Trying to be a good sport, I exaggerated my syllables. "Well, I de-*clare*. I'm just little ol' June Bug from Ala-bamy and the Texas hill country, and I just *luuv* y'all!" We all laughed heartily together.

A GRAND SCOBEE WELCOME

After we ate, Dick and I opened a treasure chest of gifts. The guests of family and friends thought of absolutely everything a young married couple could want. We were well prepared to begin filling our home with what we needed for fun and for preparing meals, including a rolling pin from his aunt who jokingly said to me, "That's for you to bump him on his noggin if he gets out of line."

His only brother, Jim, was still in high school. I found him delightful, kind, and lots of fun. Dick's best friend, Larry, was terrific and approved Dick's choice of a bride. For me, the joy was immeasurable: I finally had a big, extended family, a home, and a feeling of belonging. I delighted in sharing their name. I was a Scobee!

Dick's mother asked me to call her Mom. Her sweet smile brightened her face as she told stories about her sons growing up and showed me pictures of teenage Dick with thick dark hair, a shy smile, and apple-red cheeks.

His father told football stories about Dick and described the fun they had fishing in Green River. They showed me childhood mementos and grade cards with nearly all A's. "Our Dick doesn't like for anyone to make a fuss over him," said his father. "He kept a low profile and covered his face with his hand when they tried to take pictures of him. He sure loves his mom. When he got his first paycheck, he cashed it, then bought presents for mother and his younger brother."

I loved learning more about my new husband. The Christmas that Dick was three, his Aunt Tene gave him a toy plane with pedals like those of a tricycle. It was his favorite toy, his parents told me, bringing him hours of joy and merriment until he wore out the wheels. Then his dad made the plane into a swing and hung it from a tall cherry tree in their backyard. Soaring higher, Dick was a little closer to the sky where his child-sized hand could reach toward a star.

During his teenage years, Dick's sights were still turned toward the stars. His love of airplanes and dreams of flying were his strongest motivation throughout his school years. He sketched airplanes on his notebook paper during class time and built plastic and wooden models after school to hang on string from his bedroom ceiling. When Dick showed me around during that first day with his family, he wanted me to see his collection of airplane models even before any prized high school trophies.

That night when we were cuddled in bed, Dick asked, "What do you think of my family?"

I thought about the day, the family fun, and Dick's joy and thoughtfulness to introduce me as his wife, Mrs. Scobee, then answered, "If love had a sound, it would be of a girl's heart beating just like mine."

A BROKEN HEART

Within days after the wedding and honeymoon in Washington with his family, Dick and I returned to San Antonio and found Mother and my brothers preparing to return with Daddy to Florida. Happy but very concerned, I cried to see them all go, especially five-year-old Lee, whom I had practically raised myself. Dick held me close in his arms, assuring me that they would be all right and that we would make plans to visit them one day soon.

I didn't know just how soon. Within weeks, Daddy died suddenly. Dick took emergency leave so we could go to Florida to help with funeral arrangements. There, Mother told me that she and Daddy had argued one afternoon before going to the beach. By the shore, he had taken out his fishing pole, cast it into the surf, and then suddenly grabbed his chest and fallen over. My brothers had witnessed it.

At the hospital, doctors said he had died from a massive heart attack.

Later, when we gathered in Daddy's nicely furnished home to visit with his friends and neighbors, share stories about him, and eat the meals they brought for us, I slipped away to see the bedrooms. One was decorated with pink and white organza bedding and curtains. The feminine furniture included a desk. Inside a drawer, I found a Bible and opened it. Miraculously, my eyes fell on the well-known passage in Ecclesiastes 3: "To everything there is a season, and a time to every purpose under the heaven: A time to be born, and a time to die . . . ." In that pretty bedroom, delightfully created just for me, Daddy had placed a Bible in the drawer, perhaps remembering how I had always packed my little New Testament each time we moved.

"Oh, God," I whispered, sitting at the desk and burying my head in my arms.

I was sobbing uncontrollably when Mother came looking for me. She sat on the bed quietly at first, then spoke. "Your dad was especially proud of how he decorated your bedroom. He spent more money on this room than he did all the others combined, and he would not let anybody come in here until you saw it first."

"I broke his heart," I said, "first in Florida when he wanted me to stay and I left, and then again when I married Dick instead of coming to Florida after graduation. He created all this for me, and now I can't even thank him."

"Yes, he really loved his little girl," Mother said. "And he died doing what he loved most. He wanted to go to the beach and I didn't. In fact, I'd told him I was going back to San Antonio."

"Poor Daddy," I sobbed. "He loved *you*, Mother. He really loved you. Don't you see that's why he kept taking you back and trying to make a home for you?"

After she left me crying in my bedroom, Dick came in, kissed the top of my head, and brought me from the desk to sit with him on the bed. He wiped my tears with his neatly folded white handkerchief, and then held me in his arms. I showed him the Bible verse I'd found.

"It's one of God's mysteries," Dick whispered to me. "God was there when you needed him."

"But I wasn't there for Daddy when he needed me," I cried.

We buried Daddy the following day at the cemetery. A minister who did not know Daddy spoke words of praise. *How could he?* I wondered. *How could anybody know the sorrow Daddy knew, how hard he had worked, or how many times his heart had been broken?*

A massive heart attack. *Yes,* I thought, *he definitely died from a massive broken heart.*

TIME AND MONEY

Dick and I and my family returned to San Antonio. Mother and my brothers stayed with us for a few days, then went to Grandma's house. Finally, Mother learned she could collect Social Security payments if she could prove that she and Daddy had remarried before he died. With the monthly income, they found a house to rent, and my brothers enrolled in yet another school.

Summer eased into fall and brought sanity to our lives and our sweet home. Dick worked at Kelly Air Force Base, and I got a sales job at the Woolworth's store around the corner from our apartment. We created a budget, combining our two salaries. Together we earned nearly $200 a month. We paid $45 for rent, paid $40 for our car payment, set aside $55 for groceries and utilities, saved $10 for emergencies, and used the rest for incidentals and $10 for "entertainment."

After a late November payday, Dick told me that a great movie was playing at our favorite theater, the Majestic. We went to see *Ben-Hur*, starring Charlton Heston and featuring a fantastic chariot race action sequence and terrific historic scenes of Rome.

Closer to Christmas, we talked more about the holidays, my family and his, but mostly we followed the news about the experimental test flights taking place in the California desert called Muroc. We recalled that years before, Chuck Yeager had broken the sound barrier in the Bell X-1, and now other test pilots were lining up to fly the X-15. Dick explained, "Each test pilot is trying to go faster and higher than the one before, 'pushing the envelope' of the aircraft." We wondered what would come next.

Daddy holding June as a toddler; Mother is holding baby brother "Bobo." Birmingham, Alabama, 1944.

Dad with June (age 5) and brothers Lonnie (age 4) and Johnny (age 1). Irving, Texas, 1947.

June (age 4) and her beloved maternal grandfather, Papa Perry. Irving, Texas, 1946.

Young Lee (age 1½) on a car bumper. Odenville, Alabama, 1955.

Dick Scobee (age 3) in the pedal
plane he rode around the yard until
the wheels fell off. Auburn,
Washington, 1942.

Dick and Jim Scobee, ages 6 and 4.
1945.

Dick (age 18) is wearing his Auburn letterman's sweater. Auburn, Washington, 1957.

Dick carried this photo of June (age 15) in his wallet. San Antonio, Texas, 1958.

Dick and June (ages 19 and 16) wed at
Mayfield Park Baptist Church, 1959.

PART 2

SILVER LININGS

Chapter 8

RISING SUN

DECISIONS IN EARLY MARRIAGE

During our first Christmas holidays together, Dick and I set aside time to talk about the future. We had several serious conversations about our dreams and wishes. I was merely seventeen, and Dick had celebrated his twentieth birthday. For Christmas gifts to each other, we handwrote notes about promises and goals—not only for ourselves, but also those we wanted for each other. We wrapped the notes in scraps of Christmas paper, and then tied them like scrolls with ribbon.

At a sidewalk sale at the grocery store around the corner, we found the perfect Christmas tree, two feet tall. After paying the cashier, Dick carried the little tree home across his shoulder, reminding me of the famous Christmas scenes depicted in Rockwell art. We set it on a small lamp table, and then stood it in our front window and draped it with silver tinsel. We decorated the tree with the colorful red, green, and blue scrolls.

On Christmas Eve after attending our neighborhood church, we each took turns reading the notes we'd written: "I promise to say 'I love you' at least once a day." "I promise always to hold hands and skip and keep my little girl spirit." "I promise never to go to sleep angry at you."

To each other's surprise and delight, we had both written a goal for each other to sign up for spring semester classes at San Antonio Community College. We could only afford for Dick to take two classes, but he insisted that I take at least one class as well. We used money from our entertainment budget to pay for the courses, losing a few friends in the process because we had neither the funds nor the time to go out with them. Though enrolled in separate classes, Dick and I studied together. He focused on English and composition classes, and I satisfied my passion for math and science.

In 1960, we talked again about our financial circumstances. I suggested that Dick double up on his course work so he could graduate sooner and find a better-paying job. As is often true today, a man could often make much more money in a job or career than a woman could.

Dick protested, but we both eventually decided to get him through school while also starting our family. I finished the semester with the highest average in my algebra class. Dick circled my "A" with a red ink pen, then secured the grade report to the refrigerator with a magnet. With child-like enthusiasm, we celebrated even the smallest accomplishments at the soda fountain or ice cream shop.

During the late spring and summer, I missed many events and outings due to morning sickness. Hormones took over my body with a vengeance, but in spite of the changes, I was overjoyed that I would soon be the mother of our first child.

Dick doubled up on his college course work at night school, working weekends, and then went on the night shift with the Air Force so he could attend classes full time during the day to take the ones not offered at night.

A PRECIOUS BLESSING

Early one January morning in 1961, after a long two days of labor, our baby girl was born. Dick brought me a dozen red roses, placed them at my bedside table, and kissed the baby and me. Our little girl had a full head of dark hair and eyes as blue as the cobalt sky. Dick was one proud daddy as he held his daughter.

When the nurse came to register the birth, she asked for the baby's name. Dick looked at me to provide the explanation. I told the nurse about the pact with my dear friend years ago. "Her name will be Kathie," I announced. "She will be our beautiful Kathie Scobee."

With that pronouncement, little Kathie stretched in her daddy's arms. Dick looked at me and concluded, "I believe she likes her name and her mommy and daddy."

I asked Dick, "Now, what do you think? Is she worth all my days of morning sickness, insecurities, and irritability?"

"It's worth it all to be this little girl's daddy," he assured me. "And you?"

"We are a family now," I said, "and nothing and no one can ever separate the love we feel for each other and our children."

Dick passed the little bundle back to me. As I pulled back the edge of the blanket to look into my baby's face, I smiled. "I've never been happier in my entire life. This is joy beyond what words can express."

GAINS IN SPACE

We took Kathie home and slowly learned to be a family of three. All around us, breaking news about the space program focused on chimpanzees flying aboard rockets. Then, in April, Cosmonaut Yuri Gagarin of the USSR became the first human in space, stunning the United States with yet another step ahead in the space race.

Newspapers that month reported that the United States was involved in the Bay of Pigs invasion, an unsuccessful attempt to overthrow the government of Cuban dictator Fidel Castro. Dick grew even more immersed in his military duties, working long hours. At about the same time, he was promoted to sergeant first class, which added an extra stripe to his sleeve and more income to our growing family.

With the increase in pay, we chose to move to a nearby suburban neighborhood to rent a house with more space and a laundry room with a washer and dryer. Also, we figured my brothers could visit or even live with us if the need arose.

On May 2, 1961, the director of the Mercury project announced that Alan Shepard, Jr., would be the first of the seven (and the first American) to fly in space. On May 5, 1961 (just twenty-three days after Yuri Gagarin of the Soviet Union became the first human in space), Shepard piloted *Mercury 3* on a suborbital flight for fifteen minutes and splashed down in the Atlantic Ocean. On his return to Earth, cheers and parades welcomed Shepard, and he renamed the *Mercury 3* the *Freedom 7* in honor of all seven *Mercury* astronauts. (The names of the *Mercury* flights all ended in the number 7 to signify the bond of the *Mercury* astronauts.)

Also in May 1961, President John F. Kennedy spoke to Congress and delivered a challenge about manned space flight. He said in part, "While we cannot guarantee that we shall one day be first we can guarantee that any failure to make this effort will make us last. . . . I believe that this nation should commit itself to achieving the goal, before the decade is out, of landing a man on the moon and returning him safely to the Earth."[1] The Apollo program to land a man on the moon was born. When asked why he set this goal, Kennedy talked about the great British explorer George Mallory's climb to Mount Everest and Mallory's response to why he wanted to climb it: "Because it is there." Kennedy continued, "Well, space is there, and we're going to climb it, and the moon and the planets are there, and new hopes for knowledge and peace are there."

As Dick and I made plans for our future in early 1962, NASA sent Marine Lt. Colonel John Glenn into space. On February 20, 1962, Glenn orbited the Earth three times on *Friendship 7*. With his successful mission and many television interviews, Glenn became an instant hero. Dick admired his serious candor.

BABY STEPS TO COLLEGE

When baby Kathie began walking, I chose to return to college for at least one class per semester. Teenagers living in the neighborhood loved our Kathie and wanted to baby-sit, and we decided to invest in late afternoon help so I could attend a class with Dick. That's when I learned that we can accomplish goals and fulfill dreams—unless the car breaks down or someone gets sick. In our case, baby Kathie was sick and needed the love and attention that only her mother could provide. After I completed that semester, I postponed college indefinitely. Even when clouds of disappointment hung overhead, our little toddler brought rays of sunshine into our lives.

In the meantime, Dick studied ways to improve the livelihood for our family. He inquired about the Officer Candidate School (OCS) and the Airman's Education and Commissioning Program and Air Force Institute of Technology (AFIT). OCS would involve him in training, then commission him as an officer. AFIT would assign him to a university to study, and after graduation in a technical field, he'd be sent for officers' training and commissioning. We discussed the advantages of both avenues and chose the lengthy process that would allow him an opportunity to get an excellent college degree in engineering before being commissioned as a U.S. Air Force officer.

On other national issues, the United States was in turmoil, facing a major cold war confrontation with the Soviet Union. After the Bay of Pigs Invasion, the USSR increased its support of Castro's Cuban regime, and in fall 1962, United States military intelligence discovered that Khrushchev secretly sent nuclear missiles to Cuba.

One night in October 1962, President Kennedy was to make a major speech about Cuba. Dick and I were both in class, so during a break, we met at the car to listen to the radio and learn how the United States military would become involved. The president publicly denounced Soviet actions. He imposed a naval blockade on Cuba and declared that any missile launched from Cuba would warrant a serious attack by the United States against the Soviet Union. It was announced that all military jobs were frozen. No one was allowed to retire after serving their twenty years or after their four-year commitments. That speech solidified the decision for Dick, committing him to more years of service and most likely a career in the Air Force.

Within a few weeks, we had personal reasons to celebrate. Dick was selected for AFIT and assigned to the University of Arizona to study aeronautical engineering—his first choice. By summer, he would have completed two years of college at San Antonio College, and we would move to Tucson where he was slotted to begin with his assigned group at the university for the fall semester to complete his remaining two years toward a bachelor of science degree.

In the meantime, Mother's problems had persisted and made keeping a regular job and caring for Lonnie, Johnny, and Lee (now nineteen, sixteen, and eight, respectively) very difficult. The four of them moved several times around San Antonio, and once Mother ran away with only Lee and lived in the car in a state park where we couldn't find them.

Political situations continued to distress Mother. The United States and Soviet Union agreed to remove the missiles from Cuba without devastation to either country, but the news of nuclear missiles aimed at the United States upset Mother deeply, and no amount of consolation helped. She returned to the hospital, and my brothers came to live with us for a time. When Dick received his military assignment to go to Tucson for the university program, the three of us moved into their home until Mother could return from the hospital. Within weeks, she recovered and agreed to take her medicine, so Dick and I felt relieved about leaving my brothers for our impending move.

TAKING OFF ON A NEW PATH

In early summer 1963, Tucson welcomed us with sunshine, glorious clear skies, and crimson sunsets. While we searched for rental property, the Russians flew the first woman cosmonaut, Valentina Tereshkova, into space, and the U.S. media expressed frustration that NASA's road to the moon was strewn with obstacles.

Dick was promoted to staff sergeant and added another stripe to his uniform. He was assigned to serve with a group of other enlisted airmen as well as officers seeking master's degrees. We rented a house on Mountain Avenue, only a few blocks from the university, so Dick could walk to and from campus. He delighted in most courses, but also seriously struggled in several of his classes. To keep from waking me, he sometimes slept on the living room sofa, setting the alarm to wake during the night for extra hours of study.

Nearly all the AFIT students in Dick's group were married. We shared so much in common that friendships came easily. Frequently, I met with the other spouses to shop, take our children to the park, or play bridge.

Dick and I bought our first television in summer 1963, just in time to see the 1960s unfolding with the civil rights movement, riots and sit-ins, protests and flower children, all very foreign to a couple struggling to survive and get educated.

A FALLEN STAR AND ANOTHER BLESSING

One day in late November 1963, as I changed television channels to find entertainment for Kathie, a major news bulletin flashed across the screen. The thirty-fifth president of the United States, John F. Kennedy, had been shot while riding in a motorcade in Dallas, Texas. In disbelief, I dropped to the floor, crying with tremendous sorrow for Jackie Kennedy, who was seen scrambling in the car to get help for her husband. When we eventually heard Walter Cronkite say on the evening news that the president had died—that he was assassinated—we were in shock. Little Kathie tumbled into my arms with grave concern for her mother. She took my face into her hands and asked, "Mommy, where does it hurt?"

I explained my sadness for our country and for the president's wife and small children. Kathie was only two and a half, and I struggled with how to express my grief. "Kathie," I finally said, "the angels need the president in heaven."

"What's heaven? Where is it?"

"Heaven is where the angels live far beyond the stars."

Together, we held hands, and I prayed, "Dear God in heaven, thank you for giving us this president for a while on Earth. While he is in heaven with the angels, please take care of his family and our country. Amen."

Kathie recalled Bible stories she had learned in Sunday school and offered them for conversation. I was blessed that day to be in the presence of such curiosity and compassion from one so young.

Even with our struggles, Dick and I encouraged each other. He had much to learn and a family to care for, and I felt frustrated about living so close to a university and not being able to take classes. Still, I watched and learned from my association with the wives of the military officers, even holding office at the Officers' Wives Club and hosting lovely events that gave me confidence and brought me far from my life in Odenville, Alabama.

Television offered more up-to-date news about the space program. It had also evolved far from those early rocket days in Huntsville. NASA created the Gemini Program after realizing that an intermediate space vehicle was needed between the Mercury and Apollo projects. In April 1964, Gus Grissom won command of the first mission to fly aboard *Gemini*. He and John Young tested the new two-man capsule in orbit around Earth.

Though a wife and mother and mature for my age, I was still only twenty-one years old and excited about current events, fashion, and especially seeing the Beatles sing "I Want to Hold Your Hand" on the *Ed Sullivan Show*. That spring brought another tremendous joy into our lives when our son, Richard, was born. Dick Scobee's chest swelled as he shared even with strangers that he had a son. As parents of both a girl and a boy, we were truly ecstatic. After being reassured that we still loved her just as much, Kathie grew to love her baby brother, whom she called Richie. Dick and I were delighted with our two children, one to hold each of our hands on walks in the park. Eventually, we added a small terrier dog, Taffy, to our family, then a black and white cat we named Pyracket. We were surely the happiest family to walk on the planet.

DICK'S COLLEGE GRADUATION

As the University of Arizona graduation date neared, many graduates in Dick's aerospace class wished to be selected as Air Force pilots. Dick held out little hope to fly jets, but happily resigned himself to be an airplane maintenance officer, an important career path.

I was overjoyed that Dick's long-held dream of a college graduation was finally coming true. With his degree from the university, along with our associations and opportunities, we matured with confidence far beyond his childhood in Auburn and my youth in Odenville. Our small family of four had much to celebrate the summer of 1965. I often stopped to thank God for the prayers answered—for the opportunities that came our way, and mostly for the life we shared as a family. During those two years in Tucson, it was easy to believe in the sun even when it wasn't shining.

Note

1. "Special Message to the Congress on Urgent National Needs," President John F. Kennedy, Delivered in person before a joint session of Congress, May 25, 1961.

Chapter 9

SUNNY DAYS AND MOONLIT NIGHTS

MILITARY ASSIGNMENT IN SAN ANTONIO

Dick Scobee graduated from the University of Arizona with a bachelor of science degree in aerospace engineering on a warm spring day in May 1965. Within weeks, we packed our household goods in a moving van for storage until Dick received his military assignment and permanent orders. We traveled to San Antonio, moving our family into an apartment while Dick attended Officer Training School (OTS). Upon graduation, he would receive his commission as an Air Force lieutenant.

I was excited to be back home in San Antonio with my mother and brothers, and I especially wanted my mother to get to know her grandchildren. But when we visited her, I could tell she wasn't well. Her glazed eyes darted back and forth as the voices in her mind continued to torment her. She was angry that the United States was involved in the Vietnam War and that innocent people would die. She wanted to run away again with Lee, my youngest brother, to live in a public park so the enemy couldn't find them.

On one visit, I asked if now eleven-year-old Lee could live with us, but Mother refused. On our next visit, she tossed all the items we had given her onto the driveway. Broken dishes and glassware gift items were scattered up and down the sidewalk. When I stepped into the house to ask what had happened, she yelled, "Go away! I never want to see you again. I don't want your stuff or anything you ever gave me."

"But I don't understand why!" I pleaded.

With words that crushed me, she shouted, "It's the war! The enemy will look for your husband and find us."

That summer, I did not tell Dick about my mother's relapsed illness. He was involved in training and only saw us on weekends. I felt sad, responsible, and helpless about how to help our mother, and I was embarrassed for Dick to know about her recent tirade.

A DREAM COME TRUE

During OTS in San Antonio, Dick learned that the Air Force needed pilots and that he qualified to go to flight training at Moody Air Force Base in Valdosta, Georgia. As he often did on special occasions, Dick came home with a dozen red roses and asked me and the children, "What do you think about your daddy being a pilot for the Air Force?" Richie, barely two years old, ran around the room turning summersaults and making airplane buzzing sounds to show his excitement. Kathie, now five years old, sketched a picture of an airplane with an arrow pointing to the cockpit and the word "Daddy" printed in big letters.

Dick Scobee's dream to fly airplanes came true. We moved to Georgia, and at Moody, along with other pilots, Dick quickly advanced through the series of airplanes and instructor pilots. With each new plane he learned to fly, from the training flights in the T-34, T-37, and T-38, he grew more confident in his career choice.

The year he went through pilot training, we lived on a street around the corner from an elementary school. Each day as I drove Kathie to her kindergarten class, I yearned to be in the school teaching a class of my own students. Instead, I focused on our two precious children and looked for opportunities to make friends with the spouses of Dick's fellow student pilots.

Graduating from pilot training in 1966, Dick was near the top of his class, so he had a choice of airplanes to fly that included fighter airplanes and cargo jets. We discussed the pros and cons of each aircraft, and finally Dick decided to fly the relatively new aircraft, the C-141 cargo jet, the Starlifter. We traveled with Dick to Oklahoma for training, then to an assignment at Charleston AFB. Just north of the base, we purchased our first house. It was a new split-level home with three bedrooms (one for us and one for each of our two children). It was lovely, with all the amenities a small family could want, including a recreation room and a large backyard for the kids.

LIFE IN CHARLESTON

I have vivid memories of holidays and special family events from our first year in the house. There was no church in our relatively new subdivision, so we joined with others to build a church that we named Hillcrest Baptist Church. We loved the church family and the fellowship, and I taught the preschool Sunday school class.

Our family went on outdoor adventures, and we also enjoyed relaxing on family movie or TV night. Our favorite television show was *Star Trek*. The kids liked Captain Kirk, while Spock was Dick's favorite character. I was intrigued with the futuristic morality tales that included both good and horrid aliens. It sparked all our imaginations.

We loved to get phone calls from Dick's family and from my brothers. It was particularly good to hear from Lonnie, now twenty-one, who had a wonderful John Wayne accent. Lonnie had graduated from high school, gotten a job, and stayed at home to help with the younger boys. His hobby was drag racing at the San Antonio Drag Raceway. He drove a 1955 Chevy with a Pontiac engine and had received numerous trophies for his wins. One day, though, Lonnie called with frustration in his voice. He was concerned about our mother and needed my opinion. Uncle Gordon, who had helped many times in the past, had told Lonnie that he was now old enough to take his mother to the hospital.

Mother had told Lonnie that she was going to Charleston to take my children away from me since we lived near an Air Force base, so he called to warn me. He relayed the story that she wanted to take the children to the country, away from the city and the military, like the British did during World War II. He was concerned about her, and we agreed that she needed to see a doctor for treatments, to get more medicine, or perhaps to change her prescription.

Later, Lonnie told us how he took Mother to the hospital, explaining that at first she was reluctant to get out of the car. Then she suddenly rushed into the building, shouting to medical staff that the man following her was a criminal and ordering them to capture him. The attendants did as she said until my brother had them check the records. Once identities were made clear, Mother was taken away for a medical evaluation. After a few weeks, the psychiatrists adjusted Mother's prescription and released her from the hospital to return home after her much needed rest.

Once Mother's issue was resolved, we settled back and enjoyed life in Charleston, visiting the historic Civil War areas on the battery, discovering new tastes like she-crab soup, and going to the beach to swim in the ocean. Our dog Taffy liked to jump the waves with our children and snap at the foam. At home, our lives were divinely simple and pleasant.

It was a treat to watch Richie, now age three with red hair and freckles, playing in his miniature backyard kingdom that he created with hills, road-ways, and bridges. He built a massive maze of roads and airplane runways in the red soil under our oak tree. With his active imagination, he drove his toy dump trucks filled with sticks and acorns around a make-believe construction site.

Kathie was a kindergartner who rode the bus to and from school every day. I looked forward to greeting her each afternoon to hear about her adventures. I waited at the front door or sat on the first step of the stairs outside, eager to see her and hear about her school day. With enthusiastic gestures, she told me all about her day. They were marvelous, involved stories, and one day I asked, "Kathie, is that true? Did all that really happen?"

With her precious blue eyes, she looked up at me and admitted that some of her stories were exaggerations. When I asked why, she smiled angelically and answered, "I told those stories because they made you so happy."

That night I talked with Dick about my conversation with our daughter. "I am living out my dream to teach school through the school day stories of my child," I told him. That's when we decided it was my turn to return to my studies and fulfill my education goals. We considered the budget and decided I should begin with evening classes at Baptist College just a few miles north of our home. Dick encouraged me and was excited about getting to spend time alone with the kids a couple of nights a week.

COLLEGE FOR JUNE AND WAR FOR LONNIE

I began with the Education 101 course to ensure that I was truly on the right career path. It was intriguing enough to motivate me to attend full-time during the day the following semester. There were no scholarships, so I requested a student loan to pay the tuition. Our Kathie was in the first grade, and a church school near the college offered preschool activities for Richie.

One day during that early spring, Lonnie called us from San Antonio. "Mom's home," he said, "and taking her medicine, but you won't believe what she did to me. You know how Lyndon B. Johnson has the draft set up to enlist soldiers for the Vietnam War, and that I won't be called up because we live with our single mother?"

"Oh, no," I groaned. "What did she do?"

"Mom went to the draft board and told them they should draft me, that I'm good for nothing and probably on drugs. I go to Basic Training in May and on to Vietnam in October."

We were both concerned for the two younger brothers, but Uncle Gordon had assured Lonnie that he would watch out for them. Lonnie went on to serve in Vietnam, writing to us regularly. Sometimes the envelopes were smeared with dirt or full of red dust or grit from his location—in the jungle, the swamp, or a muddy foxhole. Once he wrote, "The other guys call me Audie Murphy, the World War II hero, because they say I know how to get around in the jungle and how to handle a gun. I guess all our backwoods experience in Odenville came in handy."

TO THE MOON AND BACK

While Lonnie served in Vietnam, Dick flew around the world on C-141 aircraft missions, and I attended college classes, the nation focused on putting a man on the moon before the Russians did. The goal of the Apollo space program, which President Kennedy began and President Johnson carried out, was to send Americans to the moon and return them home safely to Earth.

NASA selected three men—commander Gus Grissom, Edward White, and Roger Chaffee—and assigned them to fly the first *Apollo* mission in February 1967. On January 27, 1967, the three men donned their spacesuits to perform a dry run for the first manned *Apollo* flight, but before they could fly their space mission, a flash fire occurred during the training exercise, killing all three astronauts. It was a terrible setback to the nation and a tragic loss of fine people. We kept the astronauts' families in our prayers. Many feared the program would be cancelled.

MILITARY DUTY IN VIETNAM

Sadly, but not unexpectedly, within a few months we had our own personal concerns when Dick was selected for a yearlong assignment in Vietnam to fly the C-7 Caribou, a short takeoff and landing aircraft. We were prepared for him to leave us temporarily but could not help fearing the worst. Most of all, I wanted to appear supportive for our children's sakes.

Vietnam as well as *Apollo* missions and astronauts orbiting the moon made regular headline news. Though Dick and I wrote letters to each other daily while he was stationed in Vung Tau in the 535th tactical Airlift Squadron, it was often weeks before we received word and knew he was safe. Frequently, I'd see coverage of some horrific casualty on the evening news, and my thoughts always went straight to Dick. During the Vietnam War, Dick was promoted to captain and awarded the Distinguished Flying Cross, the Air Medal, and other decorations. The children and I pinned up a large wall map of Vietnam so we could follow his travels all over the country. Every day, we crossed a day off the calendar, counting down until he could come home. The children and I went to downtown Charleston and asked a professional photographer to take a photo of us to send to Dick. Dick later wrote that he kept our picture at his bedside while he wrote letters to us. Often taped inside the letter were a few coins for the kids to go to the convenience store to buy their favorite drink, most often a cherry-flavored Icee.

HAWAII R&R FROM VIETNAM

Just before Christmas 1968, the children and I joined Dick on his military leave for a Rest and Recuperation (R&R) rendezvous in Hawaii. The daytime air was temperate, and the nights were even more refreshing. With palm trees swaying in the breeze and melodic sounds of the Hawaiian ukulele, we would never have known it was Christmas except for the decorations. A large wreath hung on our grand hotel, and restaurants were decked out with bright green and red ornaments. Each night, still unaccustomed to the extreme time change, the children nodded off in our laps before they could finish their dinners. Our week together was one of the most joyous for our family, yet it was also the most heartbreaking when we had to separate for Dick to return to his duties in Vietnam. As he left me with the children in our hotel room and closed the door behind him, I leaned against the door and sobbed quietly. In a letter that came several days later, he told me how tough it was to leave us and admitted that he had actually briefly considered going AWOL.

We made the best of our time without him, traveling to Auburn, Washington, for Christmas with the Scobees. Our children especially enjoyed playing in the snow.

DICK COMES HOME

By the following April, Dick had completed his yearlong tour of duty in Vietnam and traveled home on a commercial flight. The children and I went to the Charleston airport hours early to wait. While we waited, we told each other all the things we wanted to remember to tell Dick when he stepped off the plane. Rich had lost another baby tooth and had an earache. He wore a Superman cape and could jump high off the tree house with his cape flapping in the breeze. Kathie had learned a new ballet dance to show her daddy. I was close to graduating from college. Before we knew it, the plane pulled up close to the terminal, and we all three darted for the door to greet him. We held on to him, crying happy tears and cheering. Dick noticed that we were holding up other passengers from entering the terminal. As he led us out of the way, he exclaimed, "Whoa! I need more arms to hug all my family!"

"Daddy, Daddy, you're home! You're really home!" These were the sweetest words I believe I'd ever heard. It was great to see Dick. He'd lost weight, and he looked a bit rugged and worn out, but he was home safe with us. That was all that mattered.

After a few days of vacation leave, Dick received his next assignment. He was to remain stationed in Charleston, but advance to fly a new plane, the giant C-5 Galaxy. After flight training, Captain Scobee flew the massive plane filled with cargo, military personnel, and equipment east to Europe, west to Asia, and even into the Middle Eastern countries. Sadly, he sometimes missed special family events, but we always celebrated either before or after his return.

"ONE GIANT LEAP FOR MANKIND"

In July 1969, Dick was flying the C-5 plane home from Japan when the most intriguing historic event of our age took place. The scene unfolded in black-and-white on television in front of our eyes. I woke the children and brought them to the living room in their pajamas to watch from the sofa. At only five and eight years old, they didn't understand the historical significance of the *Apollo 11* landing, but it still made a huge impression on them. The reporters showed masses of people from cultures all over the world as they huddled around televisions. They said the space journey was watched live by at least 500 to 600 million viewers, about one fifth of the Earth's population.

All day and into the night, we waited for the much-anticipated safe landing on the moon's surface. On July 20, the *Eagle* lunar module touched down on the Sea of Tranquility with Neil Armstrong and Buzz Aldrin. Armstrong radioed back the message, "The *Eagle* has landed."

The lunar module was equipped with a camera to provide live television coverage. Audiences the world over saw Neil Armstrong step on the ladder with his left foot—out of the *Eagle* and onto the moon—making his historic comment "One small step for [a] man, one giant leap for mankind." Sometime after Armstrong spoke his memorable words, Buzz Aldrin bounded out for a moonwalk. Already on the edges of our seats, the children and I eased off the sofa and onto the floor to sit closer to the television. We cheered and clapped when Aldrin planted the American flag on the moon and Armstrong took photographs of him standing next to the flag.

The sight of the two astronauts walking on the moon and leaving big boot prints in the dust had a massive impact worldwide on those who watched, especially on children who dreamed of being astronauts and on patriotic Americans who waved our flags. That day, Uncle Sam's symbolic image climbed to new heights and ushered in the era of USA bragging rights. "If we can put a man on the moon, we can do most anything."

When Dick returned home from his flight the following day, we all shared stories about the Americans walking on the moon. Though he was well aware of the historic event, he encouraged his children to tell him the details of what they saw on television, and he teased them relentlessly that the astronauts were actually walking on a moon made of cheese. More seriously, he asked, "Who remembers the words on the plaque they left on the moon?"

"What were the words? Tell us!" the children begged.

"The plaque on the moon reads, 'Here men from the planet Earth first set foot upon the moon, July 1969 A.D. We came in peace for all mankind.'"

VIETNAM STORIES

Later that summer, my brother Lonnie came to visit us. He and Dick traded stories about their military tours of duty in Vietnam. Lonnie had been seriously wounded and hospitalized in a MASH unit for weeks. He shared about what he saw while fighting at ground level, and Dick recapped what it was like to fly and experience people shooting at his plane. They told stories about Saigon and Cam Ranh Bay and the Ho Chi Minh trail. Laughing, Lonnie credited our two years in Odenville as early survival training in fending for himself.

He showed us his Purple Heart, Silver and Bronze Stars, and said he'd been recommended for the Medal of Honor for his work with the First

Cavalry Division for single-handedly wiping out a machine gun nest of enemy soldiers. Dick had stories too, but he modestly bragged about his flight crew. At Christmas, they had painted the nose of the plane to resemble Rudolph the Red-nosed Reindeer, and they'd used it to fly across the country to deliver packages and food to troops in remote areas.

Listening to my brother and my husband, I heard sadness and sorrow as they recalled their stories. I knew that they were not boasting about their experiences but were relieved to have survived the chaos of the Vietnam War. I slipped away to pray privately, expressing my grateful joy to have them both telling their stories, sitting safe and sound around our kitchen table. I also felt heart-wrenching compassion for the families whose loved ones never returned.

JUNE'S COLLEGE GRADUATION AND BEGINNING CAREER

In December 1969, I graduated from Baptist College with no special fanfare. Instead, I asked my family if we could go out to dinner to celebrate together, since they had all helped me accomplish my goal. The kids recalled how I had shared my lab experiments with them, grimacing over the tale of the animal anatomy class and the awful smell of formaldehyde that had seeped into my clothes and filled the car with the stench when I picked them up after school.

Just after the Christmas holiday, I took a teaching job in a local high school. I was finally a teacher! Ten years after graduating from high school, I had completed my college education and reached a lifelong goal. Discouragingly, my first class was full of rugged students who took advantage of my inexperience. Still, when they saw that I was not offended and could even laugh at myself, they moved away from confrontation to having fun and learning. I bubbled over with stories to tell Dick each day about the students in my classes and the joy of actually leading them to make discoveries about themselves and the world.

TEST PILOT SCHOOL IN CALIFORNIA

During the spring semester of my first year of teaching, Dick came home to tell me about an opportunity to attend the highly competitive Air Force Test Pilot School (ARPS) in the Muroc desert of California. I remembered that Chuck Yeager and other test pilots like Neil Armstrong had flown there, breaking speed and altitude records.

Dick always sought my opinion in making big decisions, hoping for my willingness to support him in his next career steps. After soul-searching discussions about life and death and opportunities, along with the thought of living in the desert, we moved our family to Edwards Air Force Base just north over the mountains from Los Angeles. To motivate our children to leave their school and friends, we told them how close we would live to Disneyland and that we would go there first and often.

We drove to what seemed like the end of the Earth and found a desert filled with cacti, Joshua trees, and miles of dry, sandy lakebeds. There in the middle of that desert, Dick attended ARPS and I taught high school.

For fun, we took the family to Los Angeles to see a movie or go shopping at the malls, and sometimes on weekends Dick and I rode motorcycles on dirt trails over desert hills, around cacti and tumbleweeds and through the air on jumps across dips. He rode a Suzuki 250, and I rode a Honda 90 that was just my size. For exercise, we jogged miles on the streets around the base during the week. After Dick graduated from the school for test pilots, he was assigned to a job testing airplanes at Edwards, so I continued to teach, and our children grew into teenagers complete with all the drama and joy of their own adventures.

A CELEBRITY ENCOUNTER

A fairly significant event took place for me while we were at Edwards. Charles Lindbergh, who had made history flying across the Atlantic in his plane, the *Spirit of St. Louis*, visited our Air Force base. We met him and his wife, Anne Morrow Lindbergh, at an event at the Officers' Club. They waited in a receiving line to meet the pilots and their spouses. As we drew near to the couple, Dick introduced himself modestly, saying, "Sir, it's my honor to meet an important pioneer in aviation." Turning to me, he said, "May I introduce my wife, June Scobee."

At first, I stood there gaping. There was so much I wanted to say, especially to Lindbergh's wife, but I could only smile and nod. Mr. Lindbergh kindly returned the smile, shook my hand, and then jokingly said, "I know you're surprised to see us. Most people think we're already dead."

"No, not at all." I grinned, finally finding my voice. "I wanted to thank you for coming, and I want to thank Mrs. Lindbergh for writing the book *Gifts from the Sea*. I often give the book as a gift to wives when they are having difficulty making changes in their lives, especially when they are hospitalized."

Mrs. Lindbergh, a petite, attractive lady, offered to talk with me later.

Before leaving the building, she stopped to thank me for my words. "What inspired you to write the book?" I asked.

"It was my therapy at a very difficult time in my life," she said. "That's how I survived the kidnapping and death of our son."

Tears welled in my eyes as I took her hand and, without thinking, reached out to hug her. When I could finally speak, we talked about her loss and what it was like to have a famous husband.

Finally, I asked for her advice to someone searching for her own identity, and she responded, "Read. Read books of philosophy." Then, as she turned to go, she slipped me a note with her address and whispered, "Write to me."

"Yes, I will. Thank you, Mrs. Lindbergh."

SKY-HIGH OPPORTUNITIES

After ARPS, Dick secured a great test flight job at Edwards flying experimental aircraft and testing their limits of altitude and speed, their engines and wings, and even the brakes on planes already in the inventory. His job also involved writing papers and giving speeches on the results of the test flights. Over the years, he logged more than 6,500 hours of flying time in 45 types of aircraft. One of the opportunities we both enjoyed was the annual Los Angeles event for the Society of Experimental Test Pilots (SETP) Conference. There we had reunions with other test pilots and their spouses, forging many long-term friendships. The people from his class and squadron, as well as our neighbors, the Sheppard family on Lindbergh Street, became lifelong family friends.

In the meantime, I reveled in the joy of being involved with teaching my students. More than working in the classroom with books and assignments, my students and I enjoyed firsthand learning experiences and off-campus activities. We studied Newton's laws of motion, then built rockets from cardboard paper towel spools, launched them, and studied their trajectories. We built airplanes and conducted experiments, changing only one variable with each flight, then recording our results by creating parabolic equations and graphs.

To get the attention of students in a science writing seminar, I rode my small motorcycle into the classroom, turned off the ignition, took off my goggles and helmet, and shook out my hair. Their writing assignment was to describe the scene. Though the administration chided me, I became infamous with the students. We took camping field trips to the desert to study the stars at night. Our host, who lived in the desert, often told stories about living under a canopy of stars. Studying the planets and constellations were important and interesting opportunities, but the greatest benefit was seeing the wide-eyed expressions of the students and their parents at the wonder of God's creation. It certainly gave us perspective and dwarfed our sense of self-importance.

I loved my students. They were motivated and interesting people. In addition to teaching high school subjects in English, Spanish, science, special classes, drama, and music, I took courses at Chapman College, completed a Master's degree, and was accepted to teach at the local college. All along, I wondered what would happen if Dick received another assignment. I had never lived anywhere in my life for longer than a couple of years, so I lectured myself to count my blessings. I felt fortunate that we had lived in the same house for nearly seven years.

Dick's ultimate test flying experience came when he was selected to fly the X-24B lifting body, the prototype plane for the shuttle. At about the same time, all of Edwards AFB eagerly watched the space shuttle *Enterprise* sitting atop a Boeing 747 on its maiden "flight" above the Mojave Desert in February 1977. Dick was so proud of his airplane experiences and his opportunity to participate with the design team and flight test of the Shuttle Carrier Aircraft (SCA).

Chapter 10

SKY-BLUE PINK BEGINNINGS

WANTED: ASTRONAUTS TO FLY SHUTTLE TRANSPORT SYSTEM

Not long after Dick flew the X-24B, he opened the *Los Angeles Times* classified section and saw an ad seeking astronauts for the space shuttle program. NASA wanted a new breed of astronaut to fly the shuttle. They not only wanted test pilots, but also scientists and engineer mission specialists, women as well as men. We both noticed that height restrictions had been lifted.

After lengthy discussions, Dick asked, "Well, what do you think? No chance they'd want an old man like me."

I encouraged Dick to apply for the new class of astronauts. "You are only thirty-seven years old! If you don't apply, you'll never know. And how many of the applicants have flown the X-24 prototype for the shuttle? You've flown the plane twice, and your test pilot friends tell me you flew it flawlessly."

Dick grinned and nodded in agreement, "Okay, if you think I should!"

"Scobee, you rascal!" I teased. "You let me talk myself into encouraging you to fly a space shuttle."

Laughing at me, he said, "Don't pack your bags just yet! A lot of great guys I know have a better chance than I do."

After the initial discussion, Dick got one surprise after another. He mailed in his application. Weeks later, he received a call to fly out to Brooks Air Force Base in San Antonio for a medical exam and other tests.

One night as we sat talking around the dinner table, Dick encouraged our children to share their thoughts about the possibility of his selection and asked them to consider how it would affect their lives. Now that they were teenagers, he wanted them to help make the decision that would bring about changes for our whole family.

Rich had been building a life-sized Burt Rutan airplane with his dad, and he rode his dirt bike in motocross races. We talked about those interests and discussed the fact that he was going into ninth grade. Rich thought he could handle the move, telling his dad how happy he'd be if Dick were selected.

We knew Kathie would find a move much more difficult. She had made great friends during her three years at Desert High School and would miss completing her senior year with them. Graciously, she hugged her dad and said, "Daddy, I'll be so happy for you if you get to fly the space shuttle." Reflecting for a minute, she looked up and asked incredulously, "That means you'll be an astronaut?" We all laughed, including their dad, who joked about himself.

Later, Dick received another call. This time he was told to fly to Houston for an interview by a panel. When he returned home from that experience, he commented about the confidence of other candidates.

"How do you think you did?" I asked.

Shrugging, he responded, "I don't know. I just answered their questions. And now I'm to write a few sentences about why I want to be an astronaut." This is what he wrote for the assignment:

> Why do I want to become an astronaut? Probably, my most compelling reasons for wanting to become an astronaut are a desire to extend and use the engineering and test pilot experience I've gained, to hopefully aid in the success of the space program, and for my own satisfaction in realizing a very longstanding personal ambition. I thoroughly enjoy being a test pilot and performing flight related tasks, and the astronaut program is, to me, a logical extension of that function into new frontiers. It is my belief that by the manned exploration and exploitation of the potentials of space and the planets, we satisfy a basic need of mankind to explore and probe the unknown, and I simply want to be an integral part of that exploration.

Not long after the interviews, NASA official George Abbey called Dick early one morning as we were heading out to work and asked, "Are you still interested in the job at NASA?"

Dick stood up at attention, grinned from ear to ear, and answered, "Yes sir, I still want the job." When he hung up and shared the news, we jumped up and down, deliriously happy and hugging each other, and then I shouted the news to the kids, who were also pleased for their dad. It seemed that no one could deserve the opportunity more than this man who rose through the Air Force ranks from an airplane mechanic to a test pilot of the shuttle prototype. When the news spread around Edwards AFB, everyone congratulated Dick. Most were genuinely delighted; others were surprised or told him he was merely lucky, but nothing dampened his spirits.

BLUNDERS ON NASA FILM

Eventually, a crew came to town to film our family doing an activity together like playing a game of chess, riding bikes, or eating a meal. It seemed rather complicated, so I asked if we could simply sit at our kitchen table and serve cake for dessert. They agreed, so I baked my favorite carrot cake covered in cream cheese frosting. I set the table with our best china and placed the cake near Dick for him to slice. While the film crew set up with several cameras and lights, the extension cords running to all outlets around the room, our teenagers talked with us about their day just as they always did.

Everything seemed fairly smooth until the film crew turned on their cameras and the lights blew a fuse. They worked a while longer, and we continued to talk more about the day. When they were set to roll with the cameras, they asked Dick to slice and serve the cake, but there was no knife! I'd forgotten to put one on the table. I was seated closest to the knife drawer, but with all the wires and cords in the way, I could not reach it, so instead I reached across the top of the cabinet to a large cleaver that I handed to Dick. It got worse. Dick tried to be funny about the knife and about serving us small pieces of cake and him a larger section. It fell flat. No one laughed. Then, when we all had bites of cake in our mouths, the fellow in charge said, "Okay, just talk to each other normally like you always do. Talk about what each of you did today."

We sat rigid, certain we had failed the film test. We had already told each other about our day. Everything seemed so artificial, forced, and phony, at which point the man asked us to tell why we wanted to move to Houston. I should have said something fairly clever about the joy of Dick getting a dream job, but instead I blurted out nervously, "Oh, Houston! After living here in the desert, it will be great to move to Houston where palm trees grow!" Silence. Everyone was stunned, and I replayed my stupidity: *Palm trees? Did I just say palm trees?* Dick said something to help us out, and then I offered the men with the cameras cake and coffee, which they seemed to welcome. Finally, the interview "activity" was over, and I was sure NASA would renege Dick's selection because his wife was so dumb.

DICK SCOBEE SELECTED TO FLY SHUTTLE

To his great surprise, though, Dick was selected from more than 8,000 applicants to participate in the group of 35 new astronauts for the class of 1978 at Johnson Space Center in Houston. Though the skies were nearly always blue in the desert, he teased our children that the skies in Houston were so happy that they were "sky-blue pink," like the mother of pearl in seashells. Kathie, Rich, and I made adjustments in our goals to support Dick eagerly in his new assignment. Kathie, who had to attend a new high school and make new friends her senior year, made the biggest sacrifice.

That summer, we drove two cars from California to Houston, one pulling a trailer that carried a partially completed home-built airplane, and the other a sports car with a motorcycle balanced on back like a hermit crab. Each car also carried a parent, a teenager, and a pet.

We bought a modest Texas-style, single-story house in Clear Lake City, a residential suburb of Houston adjacent to Johnson Space Center, and began to settle into a relatively quiet life as part of the NASA family. As Dick immersed himself in an "out-of-this-world" job, I began creating a home for us. We didn't hire an interior decorator, but we'd moved enough times that I was able to arrange furniture and hang window coverings.

ADVENTURES IN ART

I was content with my decorating skills, except for a large wall space above the fireplace mantle. The area was divided into three- or four-foot squares separated by wood paneling. No picture or painting could be centered on the space, so I decided to stitch a design to place in two parts above the fireplace. With my portable sewing machine, I cut out large peach-colored letters to represent "$E=MC^2$," the iconic sequence of symbols that represent Albert Einstein's most famous equation. I stitched them onto a background of blue sky with soft golden rays of sunshine. I measured each design to fit perfectly into the spaces, so it would appear as though the paneling was the frame for the art.

When Dick came home from work, I led him to the living room to view my creation. I anxiously awaited his approval or excitement or even critical remarks. Whether he liked them or feigned approval, he was generous with favorable comments, but the symbol led us to numerous discussions around the dinner table about the nature of clouds, mass, energy, the speed of light, time, and other phenomena.

Dick flew with the clouds and knew how powerful and even menacing they could be to a pilot, so he answered the cloud questions and generally answered the others, but he turned to me with a challenge to go after more detailed answers in a study at one of the universities in the area. In the meantime, he was inspired to create oil paintings of planes and clouds, carrying over his favorite childhood pastime of drawing airplanes. My favorite oil painting he made is of an F-4 fighter plane hovering under dark clouds but glistening with reflections from the wet wings and fuselage. Over the following months, Dick created four paintings to place in the other empty spaces on the wall on either side of my theory of relativity design. The art was a frequent reminder to us all to stretch our minds and our imaginations as we explored the world together.

VISITING MOTHER

An added advantage of living in Houston was easier visits with my mother, who lived in San Antonio and was benefiting from new medical treatments that kept her relatively mentally healthy. She had visited us in California several times and grew interested in antiquing furniture and collecting pre-WWII household items. I gave her money to assist her with her new-found hobby, and she saved some of the items for our house in Houston. Though she seemed happier than ever, Mother still harbored tremendous anger toward some of our family. After my dad died, she detested my grand-mother or chose an aunt or uncle to berate. Eventually, she began to criticize my brothers. I wondered when my turn would come.

One day, I allowed her tirades to crush me. I suggested, "Now that my children are older and Dick's career is moving along, I'd like to return to college to get a research degree."

"Why?" she asked in a rant. "Any man could do a better job than a woman."

Her words stung my pride and sense of worth for all women, but I knew better than to let her outburst hurt me for long. I thought about how a mother is supposed to encourage and be proud of her children, so I changed the subject to more domestic conversations. It was difficult to predict which topics would threaten her, so I tried to keep a low profile, hoping my children would eventually know and love their grandmother for the generous and interesting person I wanted them to remember.

FIRST CLASS OF SHUTTLE ASTRONAUTS

The group of "thirty-five new guys," or TFNG, as they dubbed themselves, bonded during their year of astronaut training. The class was divided into two groups for classroom instruction, flight training, and other activities. Often, their families were included in picnics or sporting events. Dick was designated the group leader of the blue shirts, and Rick Hauck was assigned leader of the red shirts. They actually wore T-shirts in those colors with their "TFNG" symbol emblazoned on the front. It was a clever design that depicted the orbiter in space with a cargo bay door open and thirty-five car-toon figures working and hanging all over the vehicle.

They were the first new class in ten years and the eighth in NASA's his-tory. Selected from more than 8,000 applicants, the new class shattered the mold of former astronauts. It included six women, three African Americans, and one Asian American from Hawaii. The fifteen pilots were outnumbered by twenty others: mission specialists; Ph.D.s from Harvard, Vanderbilt, and Stanford; medical doctors and surgeons with emergency room experience. Most of them told fascinating stories about themselves that intimidated me, but they told them in such a modest way that I couldn't help admiring them. Like Dick's fellow test pilots, these astronauts were serious about their work, but they knew how to have fun. Dick was happy and proud to be one of them. They all became dear friends.

STARDUSTER II AIRCRAFT ADVENTURES

We worked and studied and found time for family adventures. With another astronaut, Dick purchased an experimental airplane, an open cockpit Starduster II bi-wing plane. It was bright orange and what they called a tail dragger, because for takeoff and landing it sat back with its nose in the air, making it difficult for the pilot to see the runway. We zigzagged to see where we were going until we built up enough speed for the plane to sit itself upright and level. The nose or cowling (the cover over the engine) had been changed to resemble that of Snoopy and the Red Baron's airplanes. We wore goggles and leather helmets, and on cool days, we even wore white scarves around our necks that waved in the wind.

Dick and I went regularly to the hangar for him to work on the engine while I cleaned and polished the exterior. When we flew, Dick always gave me a turn to take off, fly straight and level, or make "S" turns. Eventually, he let me try some of the aerobatics he attempted to teach me. It was heart-thumping fun to learn how to fly—to make aerobatic lazy eights and aileron rolls in the sky, but anything that felt like a stall or spin terrified me. It wasn't long before I could actually fly without scaring myself, but I never tried to land the plane on my own, not only because I was afraid to, but also because I figured I'd always be able to fly with Dick. Someday, I reasoned, I could fly with our son Rich, too, who was eager to get his own pilot's license and join an aerobatic competition team.

HAPPILY ADJUSTED

Our children adjusted well to life in Houston. Kathie quickly found her niche at the new school, participating in the senior play and working on selecting a college to attend. She wasn't as keen on flying the airplane. Rich caught the flying "bug" learning to fly the Starduster II with Dick, then with me, after he had flying lessons. He flew other light aircraft after taking lessons from Debbie Rhiene, a champion aerobatic pilot, in exchange for doing odd jobs at her hangar. He became quite expert, eventually competing on aerobatic teams and winning trophies on a regular basis.

Nearly every night, the four of us worked at arranging our schedules so we could have dinner together. Dick and I found our teenagers so interesting that we relished conversations with them, often resorting to reference books to prove a point. We respected each other's opinions, but we sometimes required substantiation before we accepted an idea or thought.

Dick was overcome with pride for his children and told them so. He was particularly impressed and pleased with a poem Kathie wrote for him. He appreciated the imagery and symbolism about flying in space, and he wanted Kathie to know that he realized she had tremendous talent to be able to write creatively. He also gave Rich the ultimate compliment. He said, "Son, you are a great pilot. Your instincts are good. You are already a better pilot than I'll ever be." The children relished their father's love.

Chapter 11

SUNNY SKIES WITH SCATTERED SHOWERS

NEW VENTURES FOR EVERYONE

Though the local University of Houston accepted my application to teach in the College of Education, I chose instead to teach senior English at Clear Lake High School and drive north of Houston twice a week to the campus of Texas A&M University. I wanted a Ph.D. degree so I could conduct research in psychology, science, and education. During admission orientation, the head of the department told me the program was demanding and highly competitive and advised me to strive for a lesser degree. I argued that I was prepared to take the advanced subjects and challenged the professors to allow me the opportunity to major in the more rigorous program. They talked and finally agreed that I could apply for and begin the fall semester to work toward the Ph.D.

When I told Dick about the interview process, it delighted him and reminded him of our first dates as teenagers. "I've always loved your feistiness," he said. "It's intrigued me from the first day we met." His encouragement meant a lot; in fact, we took turns supporting and encouraging one another, and we worked at providing an atmosphere of support in which we could both grow.

Our Kathie graduated from high school that summer, and though she was accepted at several universities, she chose to return to California to study with her friends, beginning her course work at a community college, just as

Dick and I had done years earlier. Fiercely independent, Kathie had worked at part-time jobs after school and during the summer. We were happy to support her choice of returning to the friends she had left in California when we moved to Houston.

Rich got a landscape contract with a new housing development for several months, then he found a job at a local airport in exchange for flying instructions. The day he turned sixteen, Rich not only applied for a driver's license but also flew alone cross-country to get his airplane license. Conversation around the dinner table that night was jet-propelled with enthusiasm.

Dick also completed his year of astronaut training or Astronaut Candidate School. The thirty-five new astronaut candidates in training were dubbed the "AssCans," fondly or not so fondly, by the older or more experienced astronauts. It was intriguing to see professionals from medical, military, science, and engineering careers merge their experiences and bond as the unique first class of space shuttle astronauts. The first space vehicles were named after Greek or Roman gods, *Mercury* and *Apollo*. The new class was to fly a Space Transport System, and by April 1981, when *Columbia* first flew with the tremendous experience of both John Young and Bob Crippen, the new space program was underway and very exciting.

COLUMBIA'S FIRST FLIGHT

Dick and I and most of his classmates were among the millions of spectators at Cape Canaveral (just a few miles north of my home as a teenager in Florida). When *Columbia* lifted off the pad for the first time with its giant tank and solid rocket boosters, it was breathtaking to watch the engines fire, hear the roar crackling through the air, and feel the earth shake under our feet. Dick yelled out, "Go, baby, go!" I held my breath as it soared to the sky and disappeared into the heavens. Only then did Dick and I look to each other, wide-eyed with the stunned realization that he would accomplish the same miraculous feat some day in the not too distant future. He would fly higher, faster, and further than ever before on a rocket that lifted off the Earth with six and a half million pounds of thrust. It was mind boggling to watch and to ponder the future.

After two days and thirty-six trips around the Earth, flying twenty-six times the speed of sound, Young and Crippen headed toward Edwards Air Force Base on a 4,000-mile glide path seven times as steep as that of a powered jet airliner. They were heading toward our former residence at Edwards in the Mojave Desert to a controlled landing on the barren landscape of the Rogers dry lakebed. Their heroic flight proved that the shuttle would give the United States routine flights to space for years to come, and the names Young and Crippen are now forever listed along with the other firsts in aerospace history.

ENCOURAGEMENT TO STAY THE COURSE

As Dick and I became more involved in our separate careers back in Houston, we had less time to devote to each other. Rather than having a negative effect, it served to enlarge both our horizons and extend our appreciation and respect for each other. Our son was a teenager preparing to attend college, aiming for admission to the Air Force Academy, and Kathie was attending a California college and considering a return to a Texas university. Dick was involved in his career, so I asked them all what they thought of me returning to the university full-time to focus on completing my Ph.D. The whole family encouraged me, forming a strong circle of love and support.

I flourished under the challenge, the professors, the classes, and the professional friendships, but after a year, I began feeling guilty that I was failing as a wife and mother. I announced to my family that I planned to write a letter of resignation to my major professor because I felt that I was neglecting them.

They were silent, but their looks to each other spoke volumes. My husband expressed their disappointment, commenting that they'd all looked forward to going back to Odenville one day and meeting the people who had teased me as a child and said I'd never amount to much. They wanted to show them what I could do!

The next day as I sat at my typewriter writing the letter, my son came in and presented me with a gift wrapped in brown paper and string. I opened it to find a long block of polished wood. Turning it over, I saw a brass nameplate attached with the engraving "Dr. June Scobee." He said, "You can't quit now because this title just won't come true." He and his sister and father were saving the plaque as a gift for me when I graduated.

Encourage is a word of Latin origin that means "to put heart into." That day, my family encouraged me with all their hearts. No words or gifts could have meant more to me than their encouragement, and in a way, it was also their own sacrifice to have their mother distracted. I was home nearly every night for the family until a summer course in which I needed to stay on campus for my residency. Though I missed my family tremendously for two or three nights a week, they bonded closely in their care for each other. Kathie was home for the summer and experimented with cooking, becoming adept at preparing meals, and Rich and his dad talked about flying and even managed to fly the Starduster II around Houston and Galveston a few times. When I returned home each week, they were eager to tell me what they had accomplished, each with his or her own version of the story.

CAREER ADVANCEMENTS AND ADVENTURES IN PARIS

The following year, in 1982, Rich (now eighteen), was accepted into the Air Force Academy, a path his father had originally wanted to take choosing the road less traveled to accomplish his goals. Because life was proving complicated for our family, and I was feeling the "empty nest syndrome," we coined the motto, "All sun a desert makes, all rain a sea of sorrow." We longed for the balance of our simple "sky-blue pink" days.

In addition to shuttle flight training at Johnson Space Center, Dick also trained to fly the NASA 747 Shuttle Carrier Aircraft (SCA) that ferried the orbiter cross-country from Edwards to Kennedy Space Center. Several pilots flew the SCA to test both the SCA and the tail cone configuration on the shuttle (a cone-shaped cover was placed over the tail of the orbiter to prevent drag or other problems with the orbiter riding piggyback).

In May 1983, Dick and others flew the B-747—with the NASA prototype shuttle *Enterprise* on top—across the country and even to Europe and the Paris Air Show. I flew on a commercial flight to meet Dick in Paris. After we both arrived, we rendezvoused at an old, charming hotel near the Sorbonne.

While Dick spent his days at the Le Bourget Airport preparing to fly around the periphery of Paris, I visited the street vendors for fresh flowers for our tiny hotel room and found lovely pastry shops and fresh markets selling everything from painted landscapes of Provence and city café scenes to bins of fresh fruits and vegetables. I purchased luscious-looking ripe pears and crisp red apples to take back to the hotel and place in a bowl next to the vase of lavender irises.

The night after his flight around Paris, Dick suggested that, rather than going to the embassy with the dignitaries, we go to dinner alone at a small, family-owned bistro someone had recommended to him. With our map in hand, we meandered along the narrow streets toward the university in search of our restaurant. When we arrived, we were welcomed by the owner and seated in a cozy corner. After we studied the menu and gave our order, Dick wanted to know about my day. I told him about climbing the Eiffel Tower— less to see the city than to watch him circle high above in the clear sky. I admitted that the most unbelievable view was of him flying around the city in a gigantic bi-wing 747 with the *Enterprise* space shuttle balanced on top. I wondered what it was like.

Dick enthusiastically told me about his flight. I sat on the edge of my chair and listened to his joy. "I actually flew the plane around the periphery of Paris!" he exclaimed. "We did it. It was unbelievable! We took the plane over the city circling around the Eiffel Tower and Notre Dame. The view was spectacular! What an amazing opportunity!"

"Did it occur to you that you and the crew were making history?" I asked.

"No, I just didn't want to screw up. The responsibility was demanding; all I could do was focus on the flight pattern, but I did get to look out the window and could even see the boats on the Seine River." Grinning, he quietly reflected on the memory of the day's experience, trying to contain his excitement as he sipped the house wine placed on our table. With a toast and a clink of our glasses, he smiled at me, took my hand across the table, and said, "Thank you for being here to share this day with me."

After we had eaten but before the candlelight dinner got more romantic, someone tossed tear gas near the restaurant door that seeped inside and stung our eyes. We paid our bill quickly, and then ran out the door only to encounter more tear gas, the result of a riot. We couldn't run against the crowd toward our hotel, so we ran with it. The following day, we read in the newspaper about university students rioting (to protest unwanted reforms and radical changes as a result of the Higher Education Act of 1983) and the masses of unsuspecting citizens caught unaware.

During my outing from the hotel the following day, I ambled along window shopping, walking toward the Luxembourg gardens. The morning air felt crisp, and the window box flowers were fragrant. On my way to the gardens, I watched a few people in the shops and cafés as they talked about the NASA SCA flight. Though they spoke French, their hands gestured toward the heavens, their arms stretched out, and then they placed one hand over the other to demonstrate the orbiter riding piggyback on top of the 747.

I grinned with pride, then looked up to the heavens past the sunny blue sky to thank God for the joy Dick and I shared.

Chapter 12

REACH FOR
THE STARS

THE REWARDS OF PERSISTENCE

In May 1983, I finally graduated with my Ph.D. degree from Texas A&M University. My husband and children lovingly surrounded me at the graduation ceremony in College Station, Texas. They took several pictures of me in various "funny face" poses, wearing my robe with hood and cap. Then I gestured to each of them to join me, wanting them included because I knew I would never have graduated without their encouragement and sacrifice.

On the drive home, Dick said aloud what I was thinking: "What a long and winding road, and what unbelievable obstacles you met along the way to achieve this goal!"

I knew I couldn't have managed without him and said the reward was in the joy of learning, though for a long time I had merely wanted to prove that I could indeed achieve such a goal.

"My life was uncomplicated compared to yours," he pointed out. He was right. I'd come a long way from that kid in Odenville who struggled with one single parent and then the other, with Mother's illness, with few resources for learning, and even with living in a foster home or as the daughter of an itinerant father, moving from one school to another dozens of times.

I told Dick I was stubborn, refusing to give up or accept that I couldn't make it.

"Not stubborn," he said. "Maybe persistent. You always believed in me even when I was ready to give up. *Persistence* is greater than talent and genius. We just hung in there."

I wondered if we weren't also *blessed* with wonderful opportunities, as if by divine intervention. Dick thought for a moment, then added, "Being at the right place at the right time sure did help."

After we returned home, Dick went to the garage to get a small artist's paintbrush and pint-sized can of black paint, then went to the front of the house. Curious, I followed him to our rural mailbox that was attached to the top of a post at the front curb of the street. As I watched, he surprised me by changing the names on our mailbox from Mr. and Mrs. to read "Mr. and Dr. Scobee." I was flabbergasted, laughing in embarrassment. I felt honored that my husband was so proud of me.

TEXAS A&M BUSINESS AND EDUCATION VENTURE AT GALVESTON

Following my graduation, I served as director of the Texas A&M Space Science program. I arranged for my students to fly with professional instructor pilots and experience weightless activity. The idea was for the pilot to fly the students up and over on a roller coaster flight pattern like the NASA KC 135 used to introduce astronauts to the feeling of being weightless. Up they flew into a parabolic curve, experiencing the exhilaration, if only for a minute, that the astronauts experienced in space.

The students were also required to analyze airplane or rocket designs, then work in teams to create their own designs out of balsa wood, paper products, and glue. When the projects were completed and designated as flight ready, the test pilots had contests, launching their rockets or flying their aircrafts while controlling all variables except for the one they were evaluating. To make the test flight even more meaningful, I introduced the challenge to fly the craft safely without breaking a raw egg placed in the cockpit. We read and discussed favorite science-fiction works, took field trips to NASA's Johnson Space Center, and invited astronomers and astronauts to make presentations and introduce us to the night sky. The students were fabulous!

The other venture I took on was creating the Texas Gifted Institute (TGI), a learning center for teaching children about computers—both creating software and repairing hardware. I hired technology-savvy college students to help provide personal attention to the students, and though the kids had fun and appreciated all they learned, I discovered there was a lot more to running a business than managing a school program.

PILOTING THE SHUTTLE TRANSPORTATION SYSTEM 41-C MISSION

The NASA Shuttle Transportation System (STS) proved to be a successful spacecraft as each new flight was assigned to larger groups of astronauts on progressively more involved missions.

Eventually, Dick was assigned to fly the shuttle in April 1984 as pilot on STS Flight 41-C, whose purpose was to rendezvous in orbit with the *Solar Max* satellite. The mission was important and involved a mission specialist leaving the orbiter to spacewalk outside in the futuristic jet-propelled Manned Maneuvering Unit (MMU) to repair the satellite. It would also be the first direct ascent trajectory for a shuttle mission.

Dick was pleased with his mission, which was originally numbered STS-13 and perhaps changed to 41-C by people with a phobia for the number 13 or because of the malfunctions on the *Apollo 13*. (Their original mission was unsuccessful, but with the tremendous teamwork and resourcefulness of the crew, the astronauts returned safely to Earth.) The NASA administrator asked the Johnson Space Center director to have someone develop a new numbering system. For flight STS 41-C, the number 4 was for launching in 1984. The 1 represented launching at the Kennedy Space Center (2 would have been a new launch site at Vandenberg), and C meant they were supposed to be the third launch in 1984 (A=first, B=second, C=third).

But the crew cleverly addressed the situation with an unofficial patch of a black cat with the original number 13 painted in green on the cat's arched back. The crew also unofficially designated themselves with t-shirts depicting "Men at Work" with hard hats representing their satellite repair mission.

As Dick and the crew trained together, I made arrangements for family and special friends to join us at the Cape for the launch. Though Dick was quarantined away from all others except his crew and wife, we arranged for our guests to meet me separately at a reception that Susie Nelson, wife of mission specialist George "Pinky" Nelson, and I hosted at a beach clubhouse at Patrick Air Force Base, located just south of the Cape beyond Cocoa Beach. Our family and friends helped us celebrate Dick's grand opportunity to fly in space. Seeing the Scobees, my three brothers (now adults), our dear friends from across the country, including the Sheppard family visiting from

England, and so many others added to the excitement and softened the rough edges of our anxiety.

The five-man crew also included Bob "Crip" Crippen as commander, and mission specialists Jim "Ox" van Hoften, T. J. Hart, and George "Pinky" Nelson. The guys shortened Scobee to "Scob."

The weeklong mission began the morning of April 6, 1984. *Challenger* launched off the pad with six and a half million pounds of thrust. The ground shook. The air crackled with a thunderous roar. Shouting with our families, I covered my mouth to muffle the squeals. Tears filled my eyes and brimmed over with joy to see them climb the sky in their space rocket. Up they went, higher and higher, as the solid rocket boosters (SRBs) separated from the sides of the fuel tank and fell back to Earth. Up into the delirious burning blue they flew beyond birds and jets and straight toward the heavens. Once the fuel tank separated and they were buoyant in orbit, I imagined them blissfully shouting exhilarated cheers of triumph.

Soon, they would begin their checklist of on-orbit activities. They had two primary objectives. The first was to deploy the Long Duration Exposure Facility (LDEF), a twelve-sided cylinder about thirty feet long that carried fifty-seven experiments by researchers from eight countries. It was to be left exposed to space and then retrieved on a later flight.

The second objective was to capture, repair, and redeploy the malfunctioning *Solar Max* satellite that was launched in 1980. Pinky flew the MMU out to the satellite and attempted to grasp it, but after three attempts, the mission failed as the satellite began tumbling on multiple axes, so Crip called off the effort. The crew was disappointed and anxious about the failed attempts. During the night, engineers from NASA's Goddard Space Center took control of the satellite by sending commands to stabilize the tumbling *Solar Max* into a slow, regular spin.

The next day, Crip and Dick maneuvered *Challenger* back to *Solar Max* so T. J. could grapple the satellite with the large remote manipulator arm and place it on a special cradle in the payload bay. After Ox and Pinky heroically repaired the satellite, they deployed it back into orbit. We could all breathe again once Pinky and Ox returned safely back into the orbiter.

Other fascinating activities included a student experiment to determine how honeybees make honeycomb cells in the microgravity environment and the use of a large IMAX camera to create the IMAX movie *The Dream Is Alive*.

The mission went off as planned, but the crew made a return landing at Edwards in California instead of the Cape because of bad weather in Florida. They landed flawlessly and brought the orbiter near a building where a crowd waited to cheer for them. Our friend Betty, a NASA secretary, draped a large banner from the building to welcome Dick. Stepping out of the orbiter, he grinned at the banner with his name in big bold letters. When all five of the *Challenger* 41-C crew mates stepped onto the ground next to the orbiter, they gathered in a circle, tossed their hands into the air, and slapped them together for a group high five.

When Dick and the crew returned to Houston, all five wives were there to meet them on the tarmac as they stepped off the NASA plane. I ran ahead of the others and jumped into Dick's arms. We were so happy to have our guys back home safely from their flight in space.

THE EXHILARATION OF SPACE FLIGHT

Driving onto our street, we met friends and reporters eager to speak with Dick, but he wanted us to enjoy time alone first so he could share with me what it was like to fly in space. We drove to our favorite restaurant located on the lake, entered the restaurant decorated with nautical artifacts, and were seated at a table near a large plate glass window. We sipped champagne and talked until the stars came out, reflecting like diamonds on the ripples of the lake. But those stars in the sky sparkled no brighter than the ones in Dick's eyes as he told me about his flight of a lifetime.

He described the launch, marveling at the thrill of flying the first direct-ascent trajectory. As he talked, he kept tucking his napkin under his plate.

I had questions about the flight, but instead I asked, "Why are you tucking your napkin like that?"

"Oh," he said, blushing. "I thought I was still weightless on the space flight." He continued telling me about the launch of the LDEF and the first-ever rescue and repair mission of the satellite for the space program.

I wanted to know how the honeybees had fared.

"It was the most curious thing to watch them," Dick explained. "At first the bees were disoriented just like we were, then they gained control by linking a chain of bees together and crawling over themselves for stability. As they stabilized, their honeycomb became more symmetrical. They even had a place—a graveyard of sorts—where they put all the bees that had died."

Next I asked about the view of Earth from the shuttle window. He explained that they had little time to gaze out, but he told me the view of our planet was grand, saying that he gained a greater respect for its vulnerability.

Then we moved to practical matters, like food, bathroom breaks, and sleep. Dick told me how he and the crew members had enjoyed playing with their food in microgravity. "We had fresh fruit and vegetables for a couple of days, and even some trail mix that we tossed into the air and grabbed with our mouths as they floated around. I peeled a banana back and twirled it like a helicopter for Ox to catch." He said his special request for a jar of peanut butter and flour tortillas turned out to be a good idea, as they rolled up with no crumbs and the spread stuck without floating away.

He said that after the first night, he slept well. "It was a strange feeling to wake up and see my own hands floating in front of my face," he said. He also commented on the amount of noise on the ship. "I heard metallic and whirring sounds like I was onboard the Star Trek *Enterprise*."

Dick relaxed then, sitting quietly for a few minutes and eating his dinner. I remembered that on special occasions, we celebrated with a photograph that we took of ourselves, so I reached into my purse, took out our small camera, and handed it to Dick. He stretched out his arm, pointed the camera back at our noses, and snapped a picture, but instead of placing it back on the table, Dick simply let go of it, letting it crash to the floor. We both turned red when the wait staff turned to look at us. Dick whispered, "Oops, I forgot I wasn't in space. That's the way we handled the big cameras, just moved them over out of the way."

Muffling my laughter over our uncharacteristic dinner manners, I finally asked a more serious question. "When President Reagan called to the crew to congratulate you, he mentioned every astronaut's name but forgot yours. Didn't that make you mad?"

He insisted that it didn't. "What was important was the mission; we got the job done! That's all that mattered. We had a great mission!"

I shared his enthusiasm and excitement, letting him know how much I appreciated that he told me about his adventure first.

Never had I seen his smile so broad or heard his stories told with such delight. Our conversation leaped from his experiences aboard the spacecraft, to the great frontiers of both the inner mind and outer space, to the need for missions in life. We pondered our sense of responsibility and wondered whether people are destined or called upon to undertake certain missions or goals. That evening, I concluded that life is probably so complex because God gives us tasks, and just when we think we've managed to complete one, he challenges us with another. We agreed that we need goals and direction in our lives, but the thrill lies not in reaching the end but in learning to enjoy the journey moment by moment along the way.

KATHIE'S COLLEGE GRADUATION

That summer, our daughter reached the end of one important journey—her mission to complete her college education. She had transferred from the California community college back to Texas A&M University, where she met and married a senior university student who served in the military corps. After graduating, he received military orders for a permanent assignment at Fort Campbell, Kentucky. Kathie went with him and continued her college education at Austin Peay University in Tennessee, with a double major in journalism and English.

When making plans for commencement exercises, the university president asked Kathie if her father would be willing to give the commencement address. Dick was never one to seek speaking engagements, but he had a soft heart for his daughter, so agreed to provide the address.

When graduation day arrived, Kathie beamed with pride that we were there with her, and especially that her dad was the speaker. We observed the graduates parading across the stage with each name called, but we were watching with tremendous pride for one student—none other than Kathie Scobee, our little girl all grown up. Just like all the other parents cheering, we were certain no other moms or dads were as pleased for their child as we were for ours.

I sat in the audience watching proudly, grinning while Dick gave a talk that paralleled space flight goals with those of students working for a degree that helped prepare and train them for a career path. "Persistence is the key to reach your goal," he said, "but once you arrive at your destination, you realize the joy was in the journey along the way."

Chapter 13

SUNSHINE ASPIRATIONS

A LONG CLIMB TO REACH THE ULTIMATE GOAL

After a childhood of fascination with airplanes and flying, Dick overcame many obstacles to reach his goal of training at Moody Air Force Base, Georgia, in 1965. Over the next twenty years, he flew more than forty-five types of aircraft and logged more than 6,500 hours of flying time as an Air Force officer, a combat pilot, a test pilot, and a NASA astronaut pilot. Once we joined our lives together, I was along for the entire exhilarating ride, watching him move from airplanes to shuttle transports to the shuttle itself.

Soon Dick would lead his own crew on a space mission. The scientific principles I learned through years of experiments with homemade rockets in the classroom would play out in real life once more. As with the last shuttle launch, these principles would involve people I knew and loved sitting in an actual rocket-powered space shuttle that would blast off into space. The ultimate dream had come true: Dick Scobee was the commander of flight 51-L.

COMMANDING THE MISSION AND LEARNING THE CREW

After Dick flew the *Challenger* STS 41-C *Solar Max* flight, at least a dozen flights were scheduled and crews named. Dick waited impatiently for another turn in space. Finally, more crew assignments were announced. Dick called me from his office and said in a serious tone, "Guess who is soon to be the lucky commander of one of the greatest crews ever?"

"Now let me think. Could it be that old guy they call 'Scob'?" I teased.

He invited me to a celebration that evening, which I jokingly accepted only on the promise of champagne. We both laughed heartily at our attempt to restrain our overwhelming joy.

The year 1985 was going to be fantastic! It began with the announcement of Dick's assignment, and then an even happier event took place in late January with the birth of Kathie's baby and our first grandson, Justin. Since we married so young, we advanced to the age of grandparenting in our early forties. Someone in the astronaut office announced Dick's new status with lighthearted banter and a name tag for his space suit that read "Grandpa."

Getting down to business with the mission, Dick prepared to train with his crew for an entire year to fly the shuttle as the commander of STS 51-L. The number 5 was designated for the year 1985, the 1 for a launch from Kennedy Space Center, and the L for the twelfth mission of the year. At first, the crew numbered only five, and the mission was to launch the Tracking and Data Relay satellites (TDRS), a communications satellite, and conduct scientific experiments.

Dick was proud of his assembled crew, and I enjoyed getting to know them and their families better. We met in our homes, on picnics, and at special celebrations and meetings.

Lorna Onizuka, wife of Ellison Onizuka, lived only a few blocks from me in Houston, actually Clear Lake City near NASA. For years before that, we had lived just around the corner from each other at Edwards Air Force Base in the California desert. At first, I only knew her and her loved ones as the Hawaiian family whom I sometimes saw biking on the street or shopping at the military grocery store. We were drawn together when NASA announced in early 1978 that our spouses were chosen as finalists in the selection of the first thirty-five astronaut candidates chosen to fly the space shuttle.

At the time, Lorna's husband, El, was a test engineer, while Dick was a test pilot. They both took their work seriously and cared about their families immensely, and they both had similar interests in the space program. Dick admired El's fun-loving spirit and outgoing personality, and he was grateful to have him on his team. Both El and Lorna were fun and comical. They had a knack for finding humor in a tense situation—often at their own expense. I adored their two teenage daughters and treasured Lorna's friendship. Her ability to make us laugh at ourselves during otherwise trying moments still endears her to me. We were blessed to have her with us. I needed her. We all needed her sense of humor to see us through the exhausting and anxious days of training and waiting for the launch.

Jane Smith was the other military wife and the mother of three children. She was a petite, blonde beauty, a vivacious lady with a lovely family. Her husband Mike was a naval test pilot who was selected after Dick and El into a later group of astronauts. Dick admired Mike's experience and ability as a pilot. Whenever Dick talked with me about his day of training in the flight simulator, he spoke of Mike with great respect and admiration. Together they felt the tremendous responsibility of commanding, flying, and safely landing the "bird" (their affectionate reference to the orbiter).

Jane was my dear friend. Both of us grew up in the South and married pilots, so we understood from a wife's point of view our husband's dreams of flying. We knew their passion for the military mission, whether it was flying a new plane in a test program, training during practice missions, or the real battle of the Vietnam War. We lost friends who never returned from those missions. We stood beside their wives and children at funerals and during moving ceremonial "flyovers," when a single plane flew out and away from a formation of planes to represent the missing pilot, a lost friend. More than a pretty lady, Jane was also compassionate and thoughtful; she was my kindred spirit.

Cheryl McNair, a young, loving mother to two children, was married to Ron McNair, a talented physicist and scientist assigned to fly on his second mission into space. Ron and Cheryl were an admirable couple. They lived only a few blocks from us in Clear Lake City. Often we met one or the other jogging on the school yard track when Dick and I went jogging. I remember visiting Cheryl, a new mother just home from the hospital with her newborn son. I recall how she beamed with the newfound love of motherhood.

I admired Cheryl for her ability to juggle first a teaching career and then a job at NASA with managing a home and family. Most of all, I treasured her for her strong Christian faith and for her beauty within and without. Cheryl taught me how to pray for the strength to see God's will done. Though she was young, she was wise in the spirit of Christian love. She shared her faith freely when we met as couples or as a group, not so much in words as in actions, graceful movements, and calm melodic words of concern for the comfort of her children.

Judy Resnik was one of the first women astronauts. STS 51-L was to be her second space flight. Like most of the nation, I knew of Judy's love for her father, a kind and professional gentleman. On Judy's first space mission, this petite beauty with dark, curly hair and sparkling midnight eyes painted a poster-board sign that read, "Hi, Dad." The space cameras beamed the sign back to Earth and into the homes of people from every land around the world. She had it all, I thought—a brilliant engineer-scientist Ph.D., a job as astronaut mission specialist, a father who loved her dearly, and striking beauty. I enjoyed knowing Judy. Sometimes, Judy and I met each other for dinner at a favorite seafood restaurant in the harbor. We laughed over stories of our youth and at our attempts at jokes.

Before her first flight, Judy was assigned to assist a newscaster on the evening news to explain the goals of an earlier flight. When I complimented her several days later for her special appearance, she quizzed me, not about how she looked or came across on the camera, but on how well she explained the mission objectives. Judy Resnik was mission oriented, not unlike the rest of the crew.

Like many of the astronauts, both Judy and Dick were humble, and when asked about their professions, they usually answered, "We're just ordinary people, doing a not-so-ordinary job, working together as a team on a mission, representing the pioneering spirit of the people of our nation."

Everyone except Mike Smith had flown in space once before on flights beginning in early 1984. Each was also skilled at specific activities planned for the mission. Judy Resnik had helped develop the giant remote manipulator arm that folded into the cargo bay. She was well trained to pick up and release into orbit the TDRS. Ron McNair was a physicist assigned for scientific experiments including crucial observations of Halley's Comet. El Onizuka, as an engineer, was in charge of deploying the communications satellite. Mike Smith's expertise as a naval aviator flying fighters that he landed on aircraft carriers made him the perfect shuttle pilot for the mission.

A NEW CREW MEMBER

Earlier in the 1980s, in an effort to help the American public better connect with its space program, NASA began to investigate several options for sending an American civilian into space onboard the space shuttle. They considered sending a journalist, an explorer, or an entertainer, but ultimately decided on sending a teacher as the first civilian in space. President Reagan made the announcement inaugurating the Teacher in Space Project on August 27, 1984. NASA made *An Announcement of Opportunity* that the media disseminated in November, and applications were accepted beginning the following month through February 1, 1985. More than 11,000 teachers applied for the chance of a lifetime to fly in space and take a global classroom of students on the ultimate field trip. From the initial pool of applicants, 114 semi-finalists were selected with two from each state and territory. A National Review Panel was tasked with interviewing the candidates in Washington, DC. The panel of judges narrowed the field to ten finalists.

Not long after the announcement of the finalists, Dick sauntered into the house one evening after work, tossed his briefcase and keys on the kitchen table, and matter-of-factly asked me, "What do you think about the teacher flying with our crew on the shuttle?"

I gushed with delight. "Oh, my goodness! It's perfect! You will be a great commander of the Teacher in Space mission. It's a culmination of both our careers, merging education with space exploration."

Reaching for a cold soft drink from the refrigerator, he pulled the tab back and took a swig, but I saw a look of apprehension cross his face. Serious as always, he confessed, "I'm concerned that the ten teacher finalists haven't been told about the risks of space flight and that the shuttle is not an ordinary commercial passenger airplane as some have implied." Pausing in thought, he looked across the table to me and asked, "Do you think I should fly up to Washington, DC, to congratulate them and let them know about my concerns?"

I suggested that Dick talk with the NASA officials, then fly to DC right away before the winner was selected. He could give them a chance to change their minds. Even so, I assured him that, if I were a finalist, I would still want to fly even with the risks. I encouraged him to go and meet the applicants. He did.

Among the finalists was a high school social studies teacher from Concord, New Hampshire. After hearing about the Teacher in Space Project while attending a social studies conference in Washington, DC, Christa McAuliffe submitted an application on the deadline.

Encouraged by her husband, Steve, to apply, she wrote, "I cannot join the space program and restart my life as an astronaut, but this opportunity to connect my abilities as an educator with my interests in history and space is a unique opportunity to fulfill my early fantasies. I watched the space program being born, and I would like to participate."

The ten finalists came to Johnson Space Center in Houston for a week of thorough medical examinations, briefings about space flight, and a series of interviews with senior NASA officials. On July 19, 1985, Vice President George Bush announced that Christa McAuliffe would be the first teacher in space. Barbara Morgan, an elementary school teacher from Idaho, was

selected as Christa's backup if something should prevent Christa from participating.

Christa's training for her space mission began the following September.
In the fall, she took a yearlong leave of absence from teaching high school
history classes to begin training for STS 51-L, a mission originally scheduled
for late 1985 that had now been moved to early 1986.

Dick and I talked seriously about helping Christa and her backup
teacher, Barbara Morgan. At first, it was a "create as you go" program to get
them assimilated into the crew, make them feel welcomed, and assign them
support for training and public relations. On the day of the teachers' arrival,
I suggested that the original five crew members each present the teachers
with an apple, but Dick wanted to know the significance of apples for teachers. "You know how it is," he explained. "NASA photographers will take the
pictures and news reporters will ask me!"

"I can't imagine anyone asking about why you gave an apple to a teacher,
but it's a legend or pourquoi tale or fable with a moral at the end," I told
him.

Shrugging his shoulders, he looked at me and grinned, the characteristic
wrinkles appearing around his sky-blue eyes. "You mean like a 'why story'?"

"Yes, exactly!" I smiled. "Why teachers get apples has a great moral to
the story: 'School is a powerful place where things change and wishes come
true.'"

At the astronaut office the following day, apples were presented and photographs taken, and with the media's attention, flight STS 51-L became
known overnight as the "Teacher in Space Mission." At first, it wasn't easy to
keep up the spirit of the crew. Some members were disgruntled, so when
reporters asked to interview Dick along with the teacher, Dick chose not to
show favoritism, insisting that they interview his entire crew or only the
teacher.

In an interview conducted after Christa's selection, Dick was asked
about the significance of taking a teacher on the trip. He told the journalist,
"My perception of the real significance for choosing a teacher is that it will
get people in this country, especially the young people, expecting to fly in
space. That's the best thing that can happen to our program. The short-term
gain is a publicity gain. The long-term gain is about expectations of the
young people in this country to fly in space; they'll expect this country to
pursue the planets."

As a member of the crew, Christa's role was to teach several lessons from
the space shuttle to America's students. She planned to conduct two fifteen-

minute "live lessons" from orbit for broadcast by the Public Broadcasting System, as well as film various demonstrations on topics such as magnetism, Newton's Law, and hydroponics in microgravity.

Her first lesson, titled "The Ultimate Field Trip," was to compare daily life on the space shuttle with that on Earth, conducting a tour of the flight deck, shuttle controls, computers, and payload bay area. On the mid-deck, where Christa would sit with El Onizuka during blastoff, she planned to show her students how astronauts sleep, eat, dress, and brush their teeth— and even how the toilets work. She was to explain the crew members' roles, describe experiments being conducted on board, and demonstrate how the preparation of food, movement, exercise, personal hygiene, and sleep was different in low Earth orbit.

MISSION PLANS

Christa often visited in our home on her own or with Barbara and Barbara's husband, Clay. We felt like adopting Christa into our family because she was such a joy and because she missed her family so much. On one of her visits, she told me she was upset with NASA's insistence of heavy dependency on the scripting of her lessons. I agreed with Christa that teachers need more flexibility in what they say, especially for spontaneity and enthusiasm.

"For example," she teased, "'Good morning, this is Christa McAuliffe, live from the *Challenger*, and I'm going to be taking you on a field trip. I'm going to start out introducing you to important members of the crew. The first one is Commander Scobee.' But what I really want to say after I describe his duties and activities is that he's one of the good guys whose wife is also a teacher." We laughed together at her embellishment. She was delightful, and I hoped the rigid, uncompromising rules would not stifle her genuine charm.

In her second lesson "Where We've Been, Where We're Going," Christa planned to use models of the Wright brothers' plane and a space station to discuss why the United States is exploring space. In addition, she was to demonstrate the advantages of manufacturing in microgravity, highlight technological advances that came from the space program, and project what the future would hold for humans in space. She also wanted to have fun with her students in "Classroom Earth." She wanted to drop Alka Seltzer in water to watch the effects of weightlessness on effervescence. She planned to take along a screwdriver, a toy car, and a billiard ball for other experiments that would be filmed in the flight, narrated by Christa on her return, and distributed later as educational films. We shared ideas as freely as teachers in a faculty lounge.

When asked about the potential response to her lessons, Christa replied, "I think it's going to be very exciting for kids to be able to turn on the TV and see the teacher teaching from space. I'm hoping that this is going to elevate the teaching profession in the eyes of the public and of those potential teachers out there, and hopefully, one of the secondary objectives of this is students are going to be looking at me and perhaps thinking of going into teaching professions."

Flight 51-L was to be the second shuttle flight of 1986. Over the remaining months, another fourteen missions were scheduled to fly. The *Challenger* shuttle's primary payload was the second of NASA's Tracking and Data Relay satellites (TDRS). Working in conjunction with the first TDRS (which was deployed in 1983 by an earlier *Challenger* mission), the new satellite was expected to provide about 85 percent real-time coverage of each orbit of a user spacecraft.

Pilot Mike Smith said, "It will give us almost global coverage for shuttle missions of the future. That's going to be a big improvement not only for the shuttle, but also for the space station when it gets up later on." The satellite was scheduled to be deployed on the first day of the flight.

The 51-L mission also included forty hours of Halley's Comet observations. NASA's Goddard Space Flight Center and the University of Colorado's Laboratory for Atmospheric and Space Physics had produced a low-cost spacecraft that could measure the ultraviolet spectrum of Halley when it was too close to the sun for other observatories to do so. The project, named Spartan-Halley, would help scientists determine how sunlight breaks down water. The data was to be saved on what was then considered a robust 500 megabytes of storage.

Mission Specialist Onizuka was also going to use a camera with an image intensifier to photograph Halley's Comet from the crew cabin. In a pre-flight interview, he told journalists, "I will have about two minutes on four different orbits to photograph Halley's Comet in both the visible and ultraviolet spectrum. The objective is to try to get this data just as the comet goes around behind the sun and starts to head back out. It's a regime where we do not have any data at the present time, so I've also been told we'll probably be the only human beings to see it at that time."

In addition to the lessons Christa was to deliver from space, she was supposed to assist in operating three student experiments carried aboard the shuttle. The experiments included a Kentucky Fried Chicken-sponsored study of chicken embryo development in space (in which recently fertilized White Leghorn chicken eggs were to be subjected to weightlessness and radiation from space); research on how microgravity affects a titanium alloy; and an experiment in crystal growth.

My enthusiasm and anticipation increased even more as I grew to love both Christa McAuliffe and Barbara Morgan. In fact, the entire crew and spouses became more excited about the mission. Soon, the crew grew to seven astronauts. In 1984, NASA had announced a new policy to corporate customers who were paying millions for NASA to launch their satellites. The company would be allowed to fly an employee who qualified to go along with their satellite.

Payload Specialist Greg Jarvis was selected from more than 600 engineer applicants from Hughes Aircraft. He was scheduled to fly on STS 51-D, got bumped off, and then was told he would fly with Dick on the teacher mission. "Poor guy," Dick told me. "He's been bounced around from one flight to another too many times." Greg was to conduct fluid dynamics experiments that would test the reactions of satellite propellants to various shuttle maneuvers and simulated spacecraft movements. I looked forward to meeting both Greg and his wife Marcia sometime in early January when he joined the crew for the final days of training.

Each member of the crew had an official individual portrait made. When Dick brought his proofs home to show me, I was impressed. "They all look great," I told him, "but how will all the girls know you are married? Your hand is hidden so they can't see your wedding ring." He laughed at my feigned concern. The following day, he brought home the photo proofs taken of the entire crew. In every photo, next to his helmet, his left hand was unnaturally extended with his gold wedding band reflecting a sparkle of light. I shrieked with both approval and embarrassment as he chuckled at his ploy.

BONDING WITH CHRISTA AND BARBARA

Dick and I arranged for the crew and their families to get together as often as our schedules permitted. I especially enjoyed the chance to talk freely with teachers Christa and Barbara. We sometimes discussed the space program or what they considered doing after the space flight. Both assumed they would return to the classroom—Barbara to her elementary school in Boise, Idaho, and Christa to her high school in Concord, New Hampshire. At the time, I taught at the University of Houston and was proud of my opportunity to help pre-teachers gain a perspective and prepare for their chosen and honorable careers. I also taught graduate classes to current teachers who wanted to advance to specializations. The three of us bragged about our students and also admitted frustration over common problems. We shared similar philosophies about the joy of teaching and the thrill of watching our students grasp new concepts. We wanted to make learning fun with activities that supported the concepts and skills we taught.

Christa was delightful and a joy to get to know. She was a public relations dream not only for NASA, but also for teachers, mothers, and leaders of scout groups and church organizations. In news conferences, she radiated enthusiasm and spoke with confidence about her assignment. She marveled about the crew, comparing their training with her own preparations. In a clever gesture, she pointed to the ceiling to demonstrate all the manuals the astronauts had studied for their flight, and touched her knees to show how high her manuals reached. She was a good sport with a charming personality. We all admired her as she learned about the rigors of space flight and of being a celebrity, yet still remained a serious member of the team.

Inspired by the diaries and journals of women on the American frontier who trudged west in wagon trains and told their stories of hardship and loss, she told us, "I want to be an ordinary person who keeps a space journal about my extraordinary opportunity." Christa inspired me with her vision. She wanted to hitch her wagon to a star and become a pioneer space traveler.

We learned from each other. She wondered how I managed to look into a camera and teach a virtual classroom of students, thinking it might help her with her lessons from space. Reflecting on my experience in teaching some of the first televised classes, I explained that my students in one of my classes were located all across South Texas, in large cities and small towns.

I told Christa how I checked attendance by showing a map and pointing to the students' locations as I said their names. At the end of the semester, I arranged for a central location at a restaurant so those who wanted to meet their fellow classmates could get together. Christa and I laughed about the map she would need to help her locate her students. "It would have to reach across the United States and around the world!" I teased.

According to Dick, Christa's favorite training activity was flying with him in the T-38. She later told me, "It was my greatest thrill so far." She flew with Dick, and Barbara flew with Mike Smith for the purpose of getting accustomed to pulling G's and flying aerobatics. "I loved the dives and rolls," Christa exclaimed.

Along with our treasured conversations about teaching, I knew Christa as a mother; Scott was nine and Caroline was just six. We worked and planned together to create Halloween costumes in Houston for her to take home to Concord. After her family flew out to see her and visited Johnson Space Center, she proudly shared anecdotes about her children's first impressions.

As a mother, I knew and understood how much she missed her family. She couldn't wait to see them during the Christmas holidays and was happy to hear that she was granted leave to fly to Concord to be with them.

THE LAST HOLIDAYS TOGETHER

Christmas 1985 was a special time for all of us. With the launch only weeks away, we were excited for the crew and our loved ones. The high level of energy made the holiday season even more festive. Before Christa left for Concord, she and Barb came to my house to enjoy dinner with the families of both the upcoming *Challenger* crew and of the *Columbia* crew that had just flown with Florida congressman Bill Nelson aboard as a private citizen. Christa and Barb offered to bring a Christmas wreath and decorated cookies. We all celebrated in grand fashion.

Christmas Eve was upon us before we knew it. I was bursting with joy. Rich had come home from the Air Force Academy for the holidays, and Kathie had brought more than gaily wrapped packages: she brought her ten-month-old baby, Justin, to see us. It was our first Christmas as grandparents, and how we relished those days together!

Excitement mounted for other reasons too. We expected special guests for Christmas Day dinner and wanted to provide them with a feast. When the smell from the oven drifted out, everyone in the family helped baste the turkey, mash the potatoes, and set the table with the good china and crystal Dick had brought home from his travels to Germany. In our off-key voices, some of us sang Christmas carols. We all told stories about our lives apart, mixing in favorite memories of our lives together.

Time slipped up on us, and before we were ready, the doorbell rang and the guests arrived. Like Santa himself, Christa's teacher backup, Barbara Morgan, and her husband Clay greeted us with presents and good wishes. We most enjoyed their friendship, a priceless gift on that Christmas day. Special guests and festive meals always encourage marvelous, cheery nostalgic conversations, but this year, the conversation turned just as often to the future and our expectations of adventure and promise.

Clay, a novelist and forest service smoke jumper, had been able to join Barb in Houston as she trained to serve as Christa's backup. He had escorted her to T-38 practice flights and evening events. Dick and I had grown to love them both, feeling as though we had always known them. On a few evenings, our philosophical discussions about the workings of the mind within and our solar system beyond took us into the wee hours of the morning. Because our lives were so busy with Dick's training and my teaching, we promised each other that our friendship would continue, and that someday soon we would visit McCall, Idaho, where they lived together in a lovely log cabin on a mountain lake—quite a contrast to our "space world" in Houston.

Dick worked through most of the holidays, preparing for his flight, but he set aside Christmas Day for family and New Year's Eve for me. It was no great celebration; Dick made time for me to interview him for an article I was writing to be published in the March 1986 issue of the National Science Teacher's magazine, *Science and Children*.

As I took notes, he provided details about flying in space. I asked, "Do you have a cat or dog, and would you want to take your pet into space?" He suggested to the young readers that he would build a carpet-covered pole across the spacecraft for the cat, and that both animals would have to wear some form of diaper. We both laughed. Someday, we planned to write a children's book together about cats and dogs in space. We pictured a cat with legs outstretched and tail fluffed in fear, anticipating the end of the free fall, and the dog floating in the weightless environment with three times as much floor to sniff. We mused about the antics and shenanigans of these two unnatural choices for our imagined story, *Pets in Space*. The laughter from our play brought balance and levity to the otherwise serious days of training and preparations.

CHRISTA'S FINAL PREPARATIONS

When Christa returned to Houston after the holidays, we visited each other on several occasions, sharing more classroom stories about our students, our concern for them as people, and our keen interest to see them reach their greatest potential. We believed in hands-on, real-world opportunities for our students that included guest speakers and field trips. Recently, she told me, she had taken her class on a field trip to learn about the judicial system in a court of law. She had also created a curriculum about pioneer women, juxtaposing the westward movement with today's pioneer women who were flying into space.

One evening I invited my college students to my home to review and comment on a final project. Christa and Barbara came too. They wanted to meet my students and congratulate them for their efforts in creating a delightful children's book about a bug that slipped aboard a shuttle flight as a stowaway and told her story of adventure and conflict to young readers. My students titled their book *The Shuttle Bug*. It was a wonderful evening and a great opportunity for my graduate students (all teachers themselves) to share their project with the teachers in space. In turn, we enjoyed hearing about the astronauts' exploits, training, and hopes for the upcoming space flight.

As a remembrance of the evening, I gave each person there a small, framed piece of art matted in pastel colors with this phrase written in calligraphy: "To Teach Is to Touch the Future." I encouraged Christa to coin the phrase written by an anonymous author for herself, and "I touch the future . . . I teach!" became her motto.

More than just to educate from a spacecraft, we thought Christa could affect the course of teachers' lives. We hoped people around the globe would recognize and honor their educators who had awakened in students the desire to learn and the courage to dream. Christa wanted to explore space, to inspire children, and to teach lessons about the space frontier. The dreams of thousands were riding on Christa's shoulders.

REFLECTIONS ON A MARRIAGE

As I grew to know and love several of the crew members and their spouses, like Christa, Barbara, and Clay, I thought more reflectively about the journey my own husband was about to take. Dick and I had been married for twenty-six years, and we were still each other's best friend. I giggled like a schoolgirl at his jokes. Some would say we were soul mates, or at least that we shared heartbeats. We treasured moments together. When time permitted, he traveled out of town with me to my speaking engagements, and I went with him to events whenever I could. If someone got carried away with his being an astronaut, he quickly made a self-effacing comment or turned the attention to me by noting that I was the one with the honorable profession. It was thoughtful of him to compliment me, but Dick was modest and at times shy, blushing uncomfortably at too much attention, so he was eager to turn the focus away from himself and shine the spotlight on others.

Both the high school and university campuses were adjacent to the NASA Johnson Space Center, a quick drive around the corner from the astronaut office. When I was teaching, Dick sometimes surprised me with a call to the school to meet him in the faculty parking lot at lunch. He'd wait for me with a to-go order of cheeseburgers and Cokes. After work, we'd jog around the school track or he'd call me to join him at the gym for a game of racket ball, but he was so good at the game that he had to play with his left hand to allow a little competition.

Our marriage was wonderful, but not perfect. Some days I was impatient or unreasonable, and there were stressful days that stretched to weeks when Dick wore a permanent scowl. He had so much on his mind that I often felt comparatively unimportant. During that fall semester when schedules were impossible, our stress levels seemed unmanageable, and our tempers flared, I decided we needed to do something. I recalled my childhood "ABC" system for finding the power of positive thinking. I figured if I could use my code words to renew positive thinking when childhood seemed so hopeless, then I could use the same technique to help me find my inner compass of optimism as an adult. Life had been so comfortable in recent years that I had taken for granted my good fortune and failed to remember the helpful guidelines I'd used as a child: (1) attitude, (2) belief, (3) commitment.

I called Dick's secretary to ask about his schedule, and together she and I planned for me to pick him up early from work and take him for an overnight trip to a Galveston beach hotel, about an hour's drive south from our home in Clear Lake City.

When Dick met me waiting in the car parked outside his office, he wore the same scowl, but a grin softened it. "What's up?" he quizzed.

"Hop in. I'm kidnapping you for the rest of the day." I explained that we'd made arrangements for him to leave early, and that everything was cleared and he was in for a surprise. As we headed south, we listened to the radio tuned to his favorite station. Within an hour, we drove into Galveston, checked into our hotel, dropped off our bags, and headed for the beach.

Walking barefoot with our toes in the sand, seagulls in flight, mist in the air, and the sound of waves crashing was more therapeutic than elaborate words or even a sincere apology. At first, we walked without talking, then we held hands and grinned at each other, still without words, until the scent of salt air, soothing sounds from the Gulf, and the wide wonder of the ocean reaching to the sky washed away our tension. We realized how insignificant our problems were and how much we needed lightness in our hearts and compassionate understanding for each other.

When the tidewaters reached our feet, we stopped to roll up our pant legs and walked along, splashing in the water's edge. Soon we broke into a run in the deeper surf, churning up bubbles and tossing water onto each other. Before we knew it, we were both wet to our waists. When I ran ahead laughing, Dick chased me and tossed me about in the big waves, and together we wrestled and rolled in the surf like children.

We left early the next morning to drive back to Houston. We both returned to work a little bleary eyed and physically exhausted, yet emotionally refreshed and ready to tackle the day head on with rekindled love and a renewed spirit. Finding our sense of humor and laughing heartily at ourselves provided a balance to the ebb and flow of our lives and lifted it to a higher plane of awareness and respect.

LEAVING FOR KENNEDY SPACE CENTER

On Monday, January 25, 1986, just weeks after our quick trip to the beach, it was time to catch the NASA passenger plane to Cape Canaveral. The professional astronauts flew ahead of us in their NASA T-38 training jets, but Christa and Greg, dressed in astronaut flight suits, were to fly along with the casually dressed wives, Lorna, Jane, Cheryl, Marcia, and I. The sun shone so brightly that January morning that we had to squint to see across to the flight hangar. I held an armful of roses to present to each lady as she arrived.

We gestured anxiously to each other, to the crowd of visitors, and to the reporters who waited to see us off. Several of the reporters had become my good friends, and one of them, John Getter, would join us later at the Cape. "Hey, June," he called, "bet you wish you were the teacher taking the shuttle flight."

"Sure do," I remember calling back to him, "and I'd take you with me too." We laughed, for we had each admitted to the other that we would take a space flight in a heartbeat, especially if NASA had a need for a grandparent. We also knew Christa was the perfect teacher for the mission. She was animated, enthusiastic, intelligent, an adorable young mother, and a special friend to all teachers.

The flight to the Cape was uneventful except for the fun we had. We were relaxed friends traveling on a journey together. We talked seriously about the logistics of the rendezvous at the Cape and other events with our loved ones. Not so seriously, we compared shoes and socks. Christa wore tennis shoes rather than the big, black brogan boots Dick usually wore. I thought that was smart.

We all gathered in the center of the plane to talk and tease one another. Most of us sat around a table and faced each other. Christa sat across the table from me, talking as she answered letters to her students and friends. She was full of promise, filled with great expectations and hope for her flight. I respected her for many reasons, but at that moment, I admired her for her spunk and thoughtfulness.

We arrived at the Cape without a glitch. Dick and his crew were already there. I ran to greet him with hugs and a kiss. Then I heard a call to Christa. It was her husband, Steve, and I couldn't wait to meet him. As I turned to welcome him into the group, Dick reminded me that Steve needed to pass his primary contact physical prior to joining the crew. (The primary contact physical checked the temperature and general health of those allowed close contact with the crew so they wouldn't take the cold or flu virus on their space flight.) Steve knew that too, so he waved hello and, standing away from the crew and spouses, he tossed a t-shirt bearing the New Hampshire state seal to Christa, who passed it on to Dick for a photograph. It was as though I had always known Steve. In the next few days, we had many opportunities to get better acquainted. I predicted we would be good friends. I liked his great sense of humor, his light-hearted teasing, and his fun-loving expressions.

At noon the next day, the adult family members who had passed their physicals joined the crew members for a picnic at on the deck of an unadorned sun-bleached house that hung out over sand dunes near the shuttle launch pad. Many met each other's extended families for the first time. Christa's parents, Grace and Ed Corrigan, met Mom and Dad Scobee. They shared in the excitement of the day and the anxious moments of waiting for the shuttle launch. Our son, Rich, talked to his grandparents and to the crew, who wanted to know how he was faring at the Air Force Academy. After visiting with their families and eating fried chicken and potato salad, the seven crew members said their good-byes to all but their wives and husband, who were asked to stay longer. I took a picture of each couple in those final quiet moments before they returned to crew quarters and their spouses returned to our hotels.

Dick called Kathie on the phone. She had seen the nurse and been diagnosed with a slight fever, so she could not attend the picnic in case she was contagious. Dick wanted to tell her how sorry he was that she hadn't been able to join us. She was feeling gloomy and wanted to hug her daddy good-bye, but instead sent her hugs and kisses through the phone lines and said something funny to make her daddy chuckle. As Dick and I left the beach house to walk on the beach, we agreed that we admired our daughter's resilience and her uncanny ability for spontaneous humor. Traditionally, it's the children who learn from their parents. In our case, our children were growing up well in spite of us, and we were learning much from them.

A FINAL WALK ALONG THE BEACH

Laughter is a great release for tension, and so are the natural sounds of the sea that drew Dick and me to the beach for a final walk in the sand. The weather was unseasonably cool, but the walk gave us the chance to be alone. Dick pulled me close to him as we stepped onto the wet, sandy beach, leaving behind only a faint track of footprints at the ocean's edge.

The salty sea breeze, the sounds of hungry seagulls, and the drifting bubbles at our feet relaxed us. Most of all, we enjoyed the steady, rhythmic sounds of the ocean waves crashing upon the land, then gently flowing back and lazily churning under and into the next waves.

We decided to walk up the coast until we could see the shuttle perched on the launch pad waiting for lift-off the next morning. On first glimpse, we stopped and gazed longingly across the sandy marshland to the shuttle cloaked in the misty air. Finally, Dick spoke. "That's home away from home for the next week for the seven of us," he said. "It's going to be crowded!"

We smiled, then turned to the sea and stared at the horizon across the Atlantic Ocean. Dick hummed a familiar tune about a sailor on the sea saying farewell, then he turned to me and sang, "for you are beautiful, and I have loved you dearly, more dearly than the spoken word can tell." We embraced. I felt his love and appreciated his tender words, clinging to them for a long while and then teasing, "You and Roger Whitaker could sing a great duet."

Dick grinned and took my hand as we turned back to return to the others waiting at the beach house.

Chapter 14

SUNLIT SILENCE

We have whole planets to explore. We have new worlds to build. We have a solar system to roam in. And if only a tiny fraction of the human race reaches out toward space, the work they do there will totally change the lives of all the billions of humans who remain on earth, just as the strivings of a handful of colonists in the new world totally changed the lives of everyone in Europe, Asia, and Africa.
—Note found in Dick's briefcase, January 28, 1986

A POSTPONED LAUNCH

Mission 51-L was originally set to launch on January 22, 1986. The launch pad was familiar yet unique: Pad 39-B, last used for the Apollo Soyuz Test Project in July 1975, had been recently modified to support the shuttle program. Mission 51-L was the first shuttle to use the historic launch site. Over the next few days, the launch date was rescheduled several times more due, among other causes, to inclement weather, high winds, and a malfunctioning indicator switch for the shuttle's hatch. Between January 22 and January 27, we (the spouses) visited the crew each evening for dinner. We talked about our day, and I shared stories of our family adventures as we waited for the launch.

On the evening of January 27, we met again in crew quarters for dinner to see our loved ones before we returned to our hotels and apartments. That night, the crew seemed agitated and anxious, but they devoted time to the spouses with congenial conversations at dinner. After we ate, Dick and I went to his room for a few quiet minutes. He asked me to tell our family and friends they might as well pack up and travel back home because the flight was most likely going to be postponed again due to the temperature. All too soon, we had to leave so our loved ones could concentrate on their schedule and flight. After good night kisses, we departed crew quarters.

READY FOR LAUNCH DAY

Early the next morning before daylight, the phone rang in our rental apartment. I switched on the light in the kitchen to find the phone near the cabinet where I had placed a vase of roses from Dick. It was still dark and unseasonably cold outside.

"Hello," I said, yawning into the phone.

"Good morning, sleepy. It's a great day to go fly," Dick said. As we talked, I caressed the velvety red rose petals. He called to tell me and our children that the launch was set in spite of the cold weather.

Confused, I protested. "Last night, you said to tell our guests to go home, that you wouldn't be launching today. Will they launch the shuttle in these freezing temperatures?"

Dick explained that the engineers had knocked off the icicles that might cause a problem. I knew the jets Dick had flight-tested gained their greatest altitude in freezing temperatures, but I was still concerned about the rockets in the cold weather.

Dick assured me that he had seen pictures of rockets blasting off in snow. "They" said it was safe. Then he said, "Well, I have to go now, sweetheart. Tell the kids I love them, and I'll see you in a week. I'll miss you. Bye, honey!" He kissed a familiar sound into the phone that I held against my ear until long after the dial tone buzz drowned out my words. "I love you too," I said.

We went our separate ways—he to leave crew quarters with his team and take the crew shuttle bus to the launch pad, and once again, I to take our bus with families and friends from Cocoa Beach to the viewing sight at the Cape.

As I rode the bus in the early-morning, still-dark hours, thoughts bounced around in my head. Who were the "they" in whom Dick placed his confidence? Were these people the same "they" who frustrated Dick about certain technical concerns? Although I was anxious, I worked hard to appear calm, allowing tears to fill my eyes only for a moment when my daughter looked at me with concern on her face.

Was Kathie's concern for her dad, for me, or for her exhaustion from caring for her infant son who suffered with an earache? The previous night, we had taken him to an emergency clinic for antibiotics to help relieve the infection. *Poor dear,* I thought. *My little girl, all grown up, holding her own baby.* Was she prepared for this flight? Had I helped her, talked to her, and prepared her for the unexpected?

What about all the other children? Counting our grandson, twelve children and teenagers were with us that morning on the bus. I felt uneasy and overwhelmingly responsible for them all. Were they informed about the risks? Had their parents talked to them? Should I say something? Already, I had talked with Christa about the safety and the risks. More importantly, Dick had talked not only to her, but to all ten of the original teacher finalists about the risks of space flight.

At 8:23 a.m., January 28, the crew left the shuttle bus and walked toward the shuttle, waving to friends and television cameras. (We were still aboard our bus, but saw video later on NASA TV.) They suited up and climbed aboard the spacecraft. The launch was delayed almost three more hours to assess ice buildup on the launch pad and to allow time for the temperature to rise and ice to melt.

Continuing on our bus ride to the Cape with those dear children, I grew even more concerned. I had told the others I would fly if given the chance. I reasoned that if it was safe enough for my husband, it was safe enough for me. I prayed for us all—for our loved ones waiting to be launched, for those of us waiting to watch the launch, for our children and for children around the world waiting for lessons to be taught from the classroom in space.

When we arrived at the Launch Control Center, the families departed the buses and were escorted into the offices traditionally set aside for the crews' immediate families. The rooms offered wide windows that looked out to the launch pad, perfect for viewing the shuttle. The families waited, drank coffee or juice, and ate fresh fruit and breakfast rolls. We knew how to wait, how to chat about anything and everything to help the anxious hours pass, especially when flights were delayed.

Some paced the floor restlessly, while others peered out the window. We watched the news announcers broadcasting on the NASA select television channel, and we thoughtfully watched our children. Some of them sketched on the blackboard, others stared out the window, and still others talked with the adults and helped entertain the toddlers playing with toy cars on the floor.

Sitting quietly and trying to appear relaxed, I smiled at the family inter-actions. Looking to the front of the room, I noticed something that startled me, though I said nothing to the others in an effort to hide my anxiety. The day before, a helium-filled, shuttle-shaped balloon had bobbed near the ceil-ing, tied to a string anchored at the television. Today, the shuttle balloon was collapsed, deflated, lying on top of the television. I considered removing it, but I knew that gesture would make me feel superstitious and bring unwanted attention. *It's not an omen*, I told myself.

It was for Steve McAuliffe and his children that I felt the most concern —not that this bright attorney and former military man couldn't manage for himself, but perhaps because I felt he was least prepared for what was about to take place, and because, in the year of crew-training, I had grown to know Steve and their children through loving stories Christa had told me.

Christa and I had the opportunity to talk with each other about the safety and the risks of space flight. The subject came up in early January when she was visiting our home. Christa asked me how safe space travel was and if I would fly. We agreed that anything worthwhile in life was a risk. There had been twenty-four shuttle flights with astronauts returning home safely. "Yes, I would fly," I told her. "Without risks, there's no new knowl-edge, no discovery, no bold adventure—all of which help the human spirit to soar." Now, all around the world, children were watching and waiting for les-sons to be taught from the classroom in space.

All seven crew members had immediate family waiting with us. Besides Steve and Marcia, Jane and Lorna and Cheryl all waited with their children to send our loved ones off into space on their long-awaited journey.

The clock ticked away each moment as though it carried a heavy burden. Finally, the long-awaited final countdown was about to begin. We picked up the babies and cameras and climbed the stairs to the rooftop view-ing area. John Casper, the astronaut office liaison officer who had been with us throughout our trip, accompanied us. He was there to assist us in case of an emergency, but more than that, he was a friend. Dick and I had known both him and his wife, Kathy, since the men had served together as test pilots at Edwards Air Force Base. It was comforting to have such a capable and dear friend with us.

The morning air was crisp, but the sun warmed our faces as we peered out toward the beach to the launch site. The postcard view took my breath away. The shuttle was beautiful against the clean blue sky that served as a magnificent backdrop to what looked, from that distance, like a small replica of the actual shuttle. It glistened white and sparkling in the light.

Grouped together on top of the building, my children and I stood with Steve McAuliffe and his children. We all posed for photographs with the shuttle in the background. As we waited, I attempted to explain what was happening.

COUNTDOWN TO LAUNCH

The clock countdown began at about 11:30 a.m., at T-minus nine minutes to launch. The temperature was still a low thirty-eight degrees, thirteen degrees below the coldest temperature at which a shuttle had ever been launched. As the long-awaited final countdown began, we all tensed, standing frozen and staring at the shuttle, waiting for it to lift off the launch pad. Having been through this before, I tried to explain what would take place as the engines roared and the boosters blasted a fiery force that lifted the heavy shuttle toward the heavens. Mesmerized and distracted by the sheer raw power of the rockets, I let my words trail off.

Looking to each other and out to the shuttle, we cheered as it lifted its precious cargo higher into the sky, flying toward the heavens. The graceful white-winged spacecraft glistened in the sunlight. It was glorious and exciting—the ultimate joy for families to see our loved ones on their way. As the shuttle rocket blast propelled them higher and higher, we shouted "Hooray!" I whispered "Go baby, go!" Trembling, I prayed, "Godspeed, *Challenger!*"

Only a few anxious moments remained until separation of the solid rocket boosters (SRBs). We watched in silence as those dear to us climbed the sky. Their craft seemed to sit atop a great flume of smoke. The floor beneath us shook with the power of millions of pounds of thrust. I imagined Dick in his calm, matter-of-fact, take-charge mode. I imagined Christa in her excitement, nervously waiting for the SRBs to separate, the engines to cut off, and the buoyant lift of weightlessness to signal their safe arrival into Earth orbit.

Seventy-three seconds later, at an altitude of 48,000 feet, the right SRB, which we later discovered was leaking flame from an O-ring at one of its joints, broke loose and slammed into the external tank.

My teenage son, Rich, lovingly and protectively put his arms around his sister and me. As I reached to help Kathie with her baby, the unspeakable happened. Standing there together, watching with the entire world, we saw *Challenger* rip apart. The SRBs went screaming off on their own separate paths, and the orbiter with our loved ones exploded in the cold blue sky. It shattered into a million pieces, along with our hearts.

Frozen in the moment, we stood silently. I looked to my son and then to Christa's husband, Steve. Our eyes met. I had no words of comfort, no explanation, but I knew something terrible had happened. In stunned silence, we looked to each other for answers, for information, for hope. Gasps for air, then whispers. "What happened?" No words came, no answers. Only our glances spoke during those next helpless minutes. I saw the pain in my children's eyes. If only I could turn back that clock to stop time, but I had no power. Steve McAuliffe's eyes caught mine again. I was helpless. What could I say?

"Oh, God! It can't be," I whispered under my breath, but alarm pierced my heart and shattered all reason. I remembered how I'd encouraged the flight and explained away the risks. My mouth was quiet, but inside I was screaming at myself and at God. My knees locked, and I couldn't walk. My body was rigid. Rich took my arm and directed his sister and me down the stairs. My foot slipped from under me. I stumbled, numb and in shock. My son took my arm and supported me, helping me descend. My legs felt numb and wobbled clumsily. I stumbled again. Finally, I spoke out loud. "What about the others?" I asked. "Who's helping the children? Oh, no! Dear God, all those children!"

I pleaded with God to help me maintain my composure. When the others joined us, I patted arms, held hands, and whispered, "They'll be rescued. They'll be all right." I wasn't prepared for this role. No one ever is. Minutes later, we were on a bus to crew quarters. I know John Casper helped us, but my memory fails here. It's all a blur, like a nightmare that you try to piece together and realize the fit isn't rational. I prayed for a miracle. "Just let them survive, dear God." My head knew they were dead; my heart did not.

At a stoplight, I turned my rigid body to the window to look out. Cars were everywhere, stopped on the streets, curbs, and sidewalks. People jumped out of cars to embrace one another, perhaps friends, maybe strangers. Some sat in their cars weeping into their hands. A wave of shock jolted through the people, across the land, and around the world.

The fifteen-minute ride to crew quarters seemed like an eternity. There was no nightmare to wake from, no answers, no husband. I wanted to be strong for the others. My thoughts careened between them and myself. My life as it had been, the path Dick and I had traveled together for twenty-six years, had reached its end, and the path that reached into the future with us at each other's side had vanished in an instant. From that moment, I knew I would be changed, a different person—alone. Dick's mission, his dreams, his life, his friends—those were all gone. They were the best of us, doing the best of things for their country, pioneers crossing the frontier into tomorrow.

TRAGIC NEWS CONFIRMED

The bus delivered us to the door of crew quarters, where we had said our good-byes the night before. Now it was where we had unknowingly said our forever farewells to our loved ones. NASA officials, medical support and friends, and later our other relatives joined us at the apartment-like complex. As we gathered together in a central area, NASA official George Abbey gave us the tragic news. "All crew members are dead. They could not have survived."

No hope, no miracle, no chance. I left the others to slip away into Dick's private room because I wanted to be alone and cry out in rage. Instead, I reached for his clothes that still hung in the closet, held them in my arms, fell to my knees, and sobbed incoherent utterances.

I wept for my husband, for our shared heartbeats that were now silenced forever. I wept for the families and friends with us, for those who worked for NASA, and for the contractors. What a tragedy! What a terrible loss to humanity.

Dazed and still sobbing for Dick and his crew, I recalled teaching the Greek tragedies and Shakespeare's plays. I had used the word "tragedy" so glibly for those lessons "A tragedy is about a series of unhappy events that usually end in a disaster," I had explained. "The heroes contend against a fate they cannot escape. It affects the people closest to the event but also sweeps across whole countries, affecting an entire nation of people." With that memory, I was reminded of my children and the others. I splashed cold water on my face, brushed back my hair, and started back to my family members.

As I turned to leave the room, I saw my husband's briefcase, reached for it, and took it out to my children and Dick's parents. It was locked. I couldn't mentally recall the combination, but my fingers could, and we opened it together. There were his personal belongings: a wallet, his keys, pictures of his family, shuttle souvenir pins, business cards, astronomy charts, flight manuals, an unsigned Valentine's card that read "For My Wife," and a scrap of paper bearing his handwriting.

My children and I read the note as though it might be a message he left for us. In a way it was. It was not the miracle I prayed for, but it was a message. It was the answer to the question *why.* Why was it so important for him to fly into space? Why was he willing to risk his life? The note was a passage taken from *Vision of the Future: The Art of Robert McCall,* written by Ben Bova, one of America's great space authors. I passed it to Dick's mother. She was holding Dick's light brown jacket, which he had given her the day before to keep her warm. Cradling the jacket in her arms, she pulled it closer to her heart and read the note written by her son's hand:

> We have whole planets to explore. We have new worlds to build. We have a solar system to roam in. And if only a tiny fraction of the human race reaches out toward space, the work they do there will totally change the lives of all the billions of humans who remain on Earth, just as the strivings of a handful of colonists in the new world totally changed the lives of everyone in Europe, Asia, and Africa.[1]

Had Dick left the note in his briefcase for us to find if something happened? Did he write it on scratch paper to use as a quote in a speech? All we'll ever know is that when we most needed a message, it was there. He left for us his dream for the world and his vision for space exploration.

INEXPRESSIBLE GRIEF

Just then, we heard the announcement that Vice President Bush and two senators had flown in from Washington, DC, to give the families their condolences. The vice president stood alongside Senator Jake Garn from Utah, who had been a pilot and flown a space mission, and Senator John Glenn from Ohio, the first astronaut to orbit the Earth during the early *Mercury* missions.

Three great men, holding back tears, somehow found words to speak for the president, for themselves, and for people across our nation to say they shared in our loss and great sorrow. We stood together, the families circling the room, facing these men who suffered with us. If it had been only the day before, the commander would have greeted and thanked these national figures. No one responded. I looked to my children and then to Jane Smith, Mike's wife, who nodded to me as if to encourage a response. What could I say? I thanked them for their kind words and, still clutching the Ben Bova passage, pleaded with them to keep space exploration alive. The memory of a simple gesture as they left stands out. Vice President Bush slipped us a small piece of paper with a home phone number and this message: "Call if you need us."

President Ronald Reagan's annual State of the Union address had been scheduled for that evening. Days earlier, NASA had proposed that he mention the *Challenger* mission by saying, "Tonight, while I am speaking to you, a young school teacher from Concord, New Hampshire, is taking us all on the ultimate field trip as she orbits the Earth as the first citizen passenger on the space shuttle." Instead, the president's address was postponed. In its place, Reagan spoke at 5:00 p.m. to a grieving nation. He took care to speak to the children who were watching the live coverage of the shuttle's launch:

> I know it is hard to understand, but sometimes painful things like this happen. It's all part of the process of exploration and discovery. It's all part of taking a chance and expanding man's horizons. The future doesn't belong to the fainthearted; it belongs to the brave. The *Challenger* crew was pulling us into the future, and we'll continue to follow them.[2]

The president concluded his remarks by quoting from "High Flight," a poem originally published in 1942. "We will never forget them," he said, "or the last time we saw them, this morning, as they prepared for their journey and waved good-bye and 'slipped the surly bonds of earth' to 'touch the face of God.'"

Notes

1. Ben Bova, *Vision of the Future: The Art of Robert McCall* (New York: Abrams, 1982).

2. For the full address, please see the appendix.

Chapter 15

DARK CLOUDS

FACING THE TRUTH

Still numb with shock, I ached with the reality that would not register. I couldn't breathe. I paced the floor. I held my children. I walked from one friend to the next, trying to soothe, but what could I do? I had no answers. I asked myself, "What now? Where do we go? How do I help my family and friends?" Doctors and nurses checked our vital signs. Dick Scobee's aging parents had elevated blood pressure. Like all of us, their hearts were broken. Dick's sweet mother repeated her words softly: "He's gone. No, no!" She whispered, holding back tears, "I can't believe it. This is not what happens on the shuttle. No!"

By nightfall, an official told us they had arranged for a NASA passenger plane to fly the families back to our homes in Houston. Other family members could return with us. Someone drove us to the plane. My children and Dick's parents and other *Challenger* families were with me. We flew at a low altitude with the cabin lights off. The only sounds were soft sobs and the moan of airplane engines, as though they too were grieving.

When our plane landed in Houston, astronaut Norm Thagard and his wife, Kirby, were waiting to meet my family. Norm was Dick's close friend and shared an office with him at work. Kirby and I were also good friends. We had both taught together in the same high school. Others were there too. They held us in their arms, drove us to our homes, and shared their homes with other family members. Astronaut Fred Gregory and his wife, Barbara, invited Dick's parents to stay with them. Where there was confusion, they brought order. They made arrangements for us and protected us.

Somehow I managed to put one foot in front of the other. I talked and feigned strength, but inside I too was dying, stunned, and uncomprehending. Worst of all, I considered myself gone—usually the nurturer, caretaker, provider, I couldn't even help my own suffering children at this time. I mourned the loss of my partner in life, best friend, and companion. I mourned the loss of myself. I was different, changed, a widow—I choked on the word.

My feelings of guilt, both irrational and justified, kept crashing through my head like a thunderbolt. Why hadn't I stopped them from flying this morning? Why hadn't I prepared the children better? I grew weary of self-recrimination. Finally, I slept.

A SURREAL MEMORIAL SERVICE

Upon waking, I opened my eyes to the familiar surroundings in my bedroom, the light filtering through the sheer curtains, the dusty rose and cream-patterned walls, my husband's desk. Where was he? Oh, God. No! Like an electrical charge shooting through me again, I was awakened to the horrible reality of the day before, of the tragedy and great loss not only for my family and me but for the other families, our friends at NASA, and our nation and its children. Helplessly, I heaved great sobs. Purposefully, I prayed for strength and forgiveness.

My son joined me, and we shared tears together. Sitting next to me, he put his arm around me and consoled me. "Mama, Mama. I'm here." At that moment, my teenage son was stronger than I was. I should have been the one to console him. His dad was his hero, not so much because he was a pilot and an astronaut, but because they shared so many wonderful memories—simple pleasures and coming of age lessons that a son learns from his father.

NASA and the Johnson Space Center (JSC) made arrangements overnight for a memorial service to be held on the JSC campus near the astronaut office building. At home, we prepared to ride to the center and meet the President Ronald Reagan and First Lady Nancy Reagan.

In a private room, President Reagan met with all of us who had suffered such a great loss. They embraced us and shared their personal grief, and they listened to our words of sorrow. Nancy took my hand and hugged me and Kathie, who was holding her baby. They expressed sympathy lovingly in our quiet surroundings, and then eloquently as they joined us at the *Challenger* Astronaut Memorial Service.

My family sat to their right; Jane Smith's family sat to their left. We were all silent, some holding small American flags, others clutching handkerchiefs, one cuddling a brown teddy bear, and another cradled on the shoulder of her mother. Music played softly, the sky was overcast, and our grief-stricken spirits showed no emotion or expectations when President Reagan rose to speak.

> The sacrifice of your loved ones has stirred the soul of our nation, and, through the pain, our hearts have been opened to a profound truth. The future is not free, the story of all human progress is one of struggle against all odds. We learned again that this America was built on heroism and noble sacrifice. It was built by men and women like our seven star voyagers, who answered a call beyond duty.

The president paid tribute to the crew members, naming each of them with respect: "Dick Scobee, Mike Smith, Judith Resnik, Ellison Onizuka, Ron McNair, Greg Jarvis, and Christa McAuliffe, your families and your country mourn your passing. We bid you good-bye, but we will never forget you."

I held as tightly to those words as I did to my composure, until the end of the ceremony when we looked to the sky to observe an aerial flyover of a missing-man formation of T-38 jets. When the symbolic void was created by one of the jets that represented the death of my husband and our friends, I cried helplessly in spite of cameras and photographers poised to capture my moment of anguish.

Many times during the symbolic flight, I had stood with friends who had lost their husbands, and now they stood with me. I suddenly realized the simple and beautiful truth of the missing-man formation. Though a crew member is missing, those who remain in flight will carry on the mission. As I held my hands to my mouth to muffle the sobs, I looked to the planes and the cloudy sky beyond and made the solemn promise to continue that mission.

MEMORIALS AND CONDOLENCES

During the following weeks, I mechanically made decisions about memorials and arranged visits. I began to acknowledge but not accept the deaths. The bodies of the crew members were buried at sea, but their spirits were still here with us—angels among us, guiding us.

Friends helped me both with my personal grief and also with my unexpected and unwelcome identity, that of the *Challenger* commander's widow. Nancy Morgan, a fellow schoolteacher and treasured friend, joined me in my home to help me keep up with the housework and the personal calls. She sat with me quietly at night after my children returned to their homes. Quiet solitude with a caring friend is a heavenly gift. My dear and long-loved family friend Pat Sheppard also flew from Salisbury, England, to be with me. I was so delighted to see her that I escaped my feelings of sorrowful grief for a few days while she was with me. Kirby Thagard answered calls from the media and began organizing a room full of sympathy cards and letters. Other dear wives of the astronauts, my special friends, helped me address thank-you notes for the thoughtful letters and beautiful flowers delivered to my home. Their fresh, lively colors lifted my spirits. It seemed that everyone in our community was grieving. Elaine Balkan, the mother of one of Rich's friends, arranged with his class at the Air Force Academy to have fresh flowers delivered every Friday for many weeks.

Flowers and cards came from local television and radio stations and reporters, from local grocery stores, from schoolteachers and students, and from our church and other churches. Later they came from both friends and strangers across the country. Often they were addressed simply to the commander's widow or to the commander's children. Before long, the letters delivered came from famous kings and presidents of other countries, from Hollywood celebrities, and from leaders of industry. I was humbled to hear from them, yet overwhelmed with their outpouring of love and concern. Among all these, though, the letters from the children still waiting for lessons from space touched my soul.

NATIONAL MEDIA REQUESTS

NASA provided security guards outside my home to discourage news reporters from surprising me with interviews. I wasn't prepared to speak to anyone. I hadn't found my voice. It was only a whisper, and I had no wits about me. I finally answered a call from a *New York Times* reporter who insisted that she was writing a historical account.

"Tell me about your husband," she said.

"He was a wonderful man," I remembered, sobbing. "He loved me and his children. And I miss him so much."

"How about something descriptive?" she pushed. "What were his hobbies? What magazines did he read?"

My mind could not respond; it felt frozen. I tried to answer but I couldn't remember even one of the magazines he liked. The names of *Popular Mechanics, Newsweek, Sports Illustrated,* and *Experimental Aircraft,* or even the ones we shared like *Aviation Week* and *National Geographic,* completely slipped my mind.

The reporter, impatient and irritated, asked, "Can't you remember *anything* about your husband?"

Grieving for my husband of twenty-six years, the man I married as a teenager, whose love lifted me out of poverty, I paused after her question, trying to gain strength. Then, with an intensity that had to be heaven sent, I spoke resolutely. "If words could bring my husband back, I would speak volumes. I have no words, only sadness beyond belief of a loss that's indescribable. I'm so sorry."

I wanted to say what I was thinking, that a part of my body and soul had been severed; that part was gone, leaving me raw and hurting, and I couldn't heal.

The other *Challenger* families and I spoke to each other nearly every day. We were all hurting. Our public loss would not allow us the time we needed to grieve. If we watched the evening news, we saw again the video of the explosion. If we tuned in to a radio station, we heard music interrupted by updated news stories. If we picked up a newspaper or walked past a magazine stand at the grocery store, pictures related to the tragedy were emblazoned on the front of each piece. Our loss was more than personal; it was a national story. We all felt responsible to help our community and our nation overcome the grief. Even Space Center Houston at large was grieving.

CREATING A TRIBUTE

I recalled my childhood "ABC" system to conquer adversity, remembering to work on my attitude, belief, and courage. I suggested to the spouses that we create our own tribute to the crew that went beyond mere statues. We could continue the education mission for them. One by one, I met with them and shared my idea for a simulated space camp similar to "Space Adventures" at the Texas A&M University Institute. The idea gave us something positive to pursue.

Within only a few days, Jane Smith called to say she had agreed to a live interview on *Good Morning America.* She begged me to join her so I could talk about the space center we wanted to create for children.

I hesitated. Spilling my grief to strangers was more than I was ready to do.

Eventually, though, hearing her sincerity and her urgent pleas for help, I agreed. We were both vulnerable from our loss and overwhelmed with our sense of duty. If she could appear on the show, then I could as well.

The following morning we drove to the NASA center and found it set up for reporters and other media. The television staff seated us, threaded wires from the microphones up our sleeves, and attached the microphones to our collars. After a sound check and a countdown, the cameras began rolling.

The host spoke with sincerity, introducing us as the widow of the commander and the widow of the pilot, and asking questions about our well-being.

Jane and I met their questions with the best answers we could give. We were full of anxiety, knowing we spoke to a nation of watching people. When the host asked about the space center for children, I relaxed, speaking freely about something dear and positive that I considered a tribute to our loved ones.

Then the host asked, "Will you tell the students about the risks?"

I froze. Images of the *Challenger* exploding shook my confidence. Tears clogged my throat.

She repeated, "Will you tell the students about the risks of space flight?"

Sitting numb and inattentive, I heard tremendous commotion. "Switch cameras! Go to the other widow! Do something! Say something! Go to commercial!"

The interview was painfully over.

Returning home to the solitude of my lonely existence that day, I reflected on the question that echoed in my brain: "Will you tell the children about the risks?" Hours went by as I sat alone. Then another heaven-sent response came to me. "Yes," I said aloud, remembering my discussions with both Dick and Christa. "I will tell the children about the risks." I recalled something I had told Christa: "Without risk, there's no new knowledge, no discovery, no bold adventure, all of which help the human spirit to soar."

THE ROAD TO RECOVERY AND REBIRTH

Nearly three months after their deaths, we learned that the bodies of the seven-member *Challenger* crew had been recovered from the sea. Disbelief, pain, and anger returned. Our loss was so public. Was there no sacred privacy? In our hearts, we had already buried our loved ones. I had rationalized that they were at peace. With this discovery, the anguish returned, along with the media at my door and on my telephone. Questions—some sensitive, some not—regrets, guilt, and sorrow returned like a giant, crashing wave, knocking me down. Drowning in my own self-pity and sorrow, I looked to my guardian angels, Barbara and Fred Gregory (wonderful friends and neighbors and a fellow astronaut family), who took me into their home and cared for me away from the public and the media. They fed me and put me to bed.

My husband was buried at sea, at peace, I cried to myself. They dug him up, along with all our sorrow, for us to bury again. I felt the open, gaping wound of grief and knew how public it was. I hated this life. I was angry with God for letting everything happen. The energy used for anger, hate, and grief began to weaken me. I longed for privacy, even for death, to see Dick again and say good-bye to this painfully public life. Joy and stability had begun the day he came into my life, and it had died when he left me.

In my irrational state, I prayed for death. Then I had a vision or a dream that was very real to me. Dick stood at my bedside, visiting me from behind a grand curtain of light of indescribable proportion, blinding but also gentle. I asked Dick if I could join him.

"Not yet," he said. "They tell me it's not your turn. You still have life to live and things you must do." His voice without form was tender. "It's wonderful here. The best part of it is you can take the love with you."

I asked how this could be, and he simply replied, "I can be anywhere—here, there with our children, anywhere in less than a flash." Then he vanished into the soft illumination of a starburst-like quality that diminished to a glow like that of a firefly, finally flickering into a void of darkness.

I cried an agonizing prayer. "Dear God, please take me. If you won't take me," I begged, "then give me strength to live this life, and help me solve these problems and overcome this guilt. Please, dear Lord in heaven!"

This experience jolted me back to the living. That moment, I became like a child again. God was in control—not me. For the first time in my life since grappling with losses in my childhood, I relinquished complete control to God. A joyous spirit challenged me to live, accept my problems, and discover the joy that awaited me in a new life. For the first time, I didn't feel alone. My faith was renewed.

The next morning, I was left alone, but I didn't feel alone. I knew God was with me and in control. The pressure, the anger, the pain, and the guilt drained from my body. A part of me had died; a stronger, more centered, and saner self was born. I stared into the mirror, and through different eyes I saw a new person. My cheeks were flushed pink like the little girl I once was.

The numbness I had known for months subsided. Piece by piece, I became more alive.

I stepped outside onto the shady, green lawn and thanked the early morning light for chasing away the night. I looked at my hands and arms. The tense, tight feeling was gone. I was more alert, more aware of my body, and more conscious of life around me. For the first time in months, I felt the tingle of a breeze floating across my skin. A ray of April sunlight fell warm across my back. Sounds consoled me—birds singing from treetop perches and children at play, laughing in the distance. A single golden daffodil bent forward as if to welcome me to the new life of spring.

A car pulled up, and my little grandson ran out. A one-year-old toddler fell pell-mell into my arms. What beauty this life held! What joy! How selfish, I thought, to want to leave this—to leave them. As I embraced my daughter, following close behind on the heels of my grandson, tears rained down my cheeks, but this time they were tears of joy for a rebirth to live out this life with my children and their children on a new journey, in a new direction, along whatever new path God unfolded for me.

FUNERALS

Funerals, services, and burials were arranged during the month of May. Each family of the seven astronauts planned according to their religion, family tradition, or the request of the loved one: Protestant, Catholic, Jewish, Buddhist; some private, others public.

I grappled with Dick's request to be buried privately and quietly in an old apple box and also with his early thoughts about cremation. I talked to our children, Dick's parents, and our pastor. We chose to honor the man who relished his privacy. Already, a part of him was left buried at the crash site, scattered at the bottom of the sea on the Florida coast near the site of Cape Canaveral that he so loved. Our children and I buried the part recovered from the wreckage quietly and privately, along with his astronaut pilot wings, in the wooden box he requested. A modest headstone, symbolic of the plain and simple life he most wanted, was placed on his gravesite near the *Challenger* Astronauts Monument at Arlington National Cemetery.

We also arranged a service for the more public figure—the astronaut and military man who loved and served his country well. Friends and family joined us for this more formal ceremony. They walked with us from the chapel to the gravesite to pay their last respects. As a bugler sounded taps, I placed flowers on Dick's grave located across from the grave of the Unknown

Soldier, on a spread of grassy green beneath shade trees that reach to the sky just beyond the Washington National Airport flight pattern.

Those were only the gravesites for Dick's body, laid to rest on May 19, 1986, forty-seven years after his birth. He is not there, though. On January 28, 1986, Dick went with his six crewmates as they climbed a golden beam of light sunward to the heavens intending to touch the stars, but instead put out their hands and "touched the face of God."

Life moved forward more certainly after they were all laid to rest. We learned lessons—not the planned lessons, but new lessons about life. We learned that life is fragile and temporary and that, even so, goals are worth striving for.

OVERCOMING ADVERSITY

People in Houston recognized me wherever I went—at the grocery store, in restaurants, at the airport. They stopped me, encouraged me, and opened their hearts to me with their stories.

Everyone meets adversity at some time in their lives, these people told me. We lose loved ones in death or divorce. We suffer great illness or sorrow. Some of us live quiet lives of desperation, seek fame only to fail, or seek fortune only to lose. Some become broken and bitter about their experiences, while others overcome and become greater because of them. The real worth in life, these people confessed to me, is loving, learning, and helping others.

What powerful messages! Those who shared with me had also known the depths of despair and the heights of great joy. They had witnessed both the calm and the turbulence of the sea. Listening to them, I wondered again if I could redirect the energy used for grief and anger into a more creative purpose.

Would the resulting joy reach the heights opposite but equal to the depths I had known? Could I achieve it by helping others?

One day, amid these thoughts, I recalled a powerful childhood memory that helped me through years of early adversity. A visitor gave my mother some tangerines and a book. The tangerines were sweet and delicious, but the book, *The Power of Positive Thinking* by Norman Vincent Peale, was what fascinated me. Even at nine years old, I could understand a lot of it. It led me to my "ABC" guide, which, along with the Bible, supported me for years. One of the book's strongest messages was that we can use positive thinking to cope with even the worst circumstances, and that God is always with us.

Putting my sorrow aside and opening my eyes to life beyond my front door, I saw a nation's can-do spirit fading. I began to watch the evening news, ask questions, and read reports. The Rogers Commission had completed its study of what happened to cause the *Challenger's* fiery explosion. The news was awful for all seven families, the NASA family, and our nation. Overconfidence, poor management, neglect, flawed O-rings, poor decisions about the limits of technology—science had jumped ahead of the collective conscience of humanity. Technology, rather than people, was leading us into the future. Something had to happen to wake us up. It did. At a terrible expense and unbelievable loss, we learned the mighty lesson that moving forward in our nation's pioneering effort requires people working together as a team, focused not only on where the technology can take us, but also on the education and conscience of the people who are taking us there.

When my husband and his six friends died aboard the *Challenger*, a piece of our nation's spirit died too. The nation's pride in its space program had spilled over into our lives, both in the workplace and the home. It kept us on the cutting edge of technology and on the frontiers of the mind and space. As a nation, we felt that if we could put a man on the moon, we could accomplish most anything.

The miracle of space travel made the symbolic chest of Uncle Sam swell with pride. Our country's space program had taken us to great heights. After falling from these heights on the day the *Challenger* exploded, we needed to recall John F. Kennedy's words from a speech he made at Rice University in Houston on September 12, 1962: "Surely the opening vistas of space promise high costs and hardships, as well as high reward."

Dick and June with Kathie (age 1). 1962.

Kathie (age 3)
and baby Richie.
Tuscn, Arizona, 1964.

Kathie and Richie riding
their Dad on Halloween.
Valdosta, Georgia, 1966.

June, Kathie, and Richie on a tandem bicycle, a favorite picture sent to their daddy, Dick Scobee, while he was in Vietnam for a year. Charleston, South Carolina, 1969.

Dick and June after her graduation from Texas A&M University, 1984.

Dick and Kathie celebrate after Kathie earns her driver's license. Edwards AFB, California, 1977.

41-C Crew. Left to right: Commander Bob "Crip" Crippen,
Mission Specialists T. J. Hart, Jim "Ox" van Hoften,
George "Pinky" Nelson, Pilot Dick "Scob" Scobee.

Shuttle cloaked in fog waiting for launch.

Christa McAuliffe in Zero-G flight, weightless experience with
Congressman Bill Nelson and Barbara Morgan (Christa's backup)

Dick flew the 747 SCA with orbiter piggy back. They flew it cross country with
the tail cone covering the engines to prevent drag. Paris, France, 1983.

SCA with orbiter on top and family San Antonio, Texas, 1983.

Dick and June kiss in front of Shuttle Carrier Aircraft (SCA).
This picture was taken during our "round-the-world-trip." Europe, 1983.

Dick and June at a party just before the 41-C flight. Houston, Texas, 1984.

Dick and June unknowingly take a final walk on the beach
in Florida on January 26, 1986.

Challenger crew with Barbara Morgan (Christa's back-up). Back row, left to right: El Onizuka, Barbara Morgan, Christa McAuliffe, Greg Jarvis, Judy Resnik. Front row: Mike Smith, Dick Scobee, Ron McNair

Christa McAuliffe—"The Write Stuff"

Crew members of Mission 51-L walk out of the Operations Checkout Building on their way to board *Challenger 7* for the launch. Kennedy Space Center, January 28, 1986.

June with President and Mrs. Reagan at the *Challenger* memorial service.
Houston, Texas, January 30, 1986.

Dick was buried at Arlington quietly and privately, along with
his astronaut pilot wings, in the wooden box he requested.
May 1986.

Arlington National Cemetery Challenger Astronaut Memorial.
Arlington, Virginia.

The White House
Washington, DC
July 2004

The President of the United States of America
Awards this

CONGRESSIONAL SPACE MEDAL OF HONOR
to
Francis R. Scobee

Francis R. Scobee distinguished himself as an astronaut, as an Air Force officer, and as the Commander of the
STS 51-L crew on board the Space Shuttle Challenger. His meritorious and dedicated service to the Nation and his
pioneering contributions to human space flight are a credit to him, to his family, and to his country.

Montage created of Dick Scobee's picture with Space Medal of Honor.

Challenger spouses singing "God Bless the U.S.A." with Lee Greenwood
at the first gala in Washington, DC. Walter Cronkite was the emcee,
and Oprah Winfrey and Vice President Bush were presenters.
Left to right: Jane, Lee, Marcia, Lorna, June, Steve, Cheryl, Chuck. 1988.

Challenger spouses Lorna, Marcia, June, Cheryl, and Jane and Judy's brother
Chuck Resnik—all Founding Directors. Washington, DC, 1996.

June with John and Annie Glenn at Kennedy Space Center.

Barbara Bush accepts the Vase Award, the first
Challenger Center Presidential Award, Washington, DC, 1996.

To: dear June Scobee Rodgers —
all good wishes, Barbara Bush

June Scobee Rodgers and Barbara Bush at the White House. April, 1989.

June and Don at a White House dinner, April 1989.

Robert McCall's first concept design for the Challenger Center. 1986.

PART 3

VAST HORIZONS

Chapter 16

CLOUDS ARE NOT FOREVER

THE CHALLENGER AS HISTORY

The spirit of exploration and discovery has always shaped America. Whether on land, in the sea, or in the air, these explorations and discoveries have forever changed the face of our nation—and often the world.

After the *Challenger* tragedy, I was frequently asked to speak to audiences about the accident and the continuation of the space program. The audience had more questions than I could answer.

"Do you fear the *Challenger* accident might be the final story in this country's history book of space exploration?" they wondered. "Will the next generation have no explorers? Will this tragedy cause a dead end on the road of exploration to space?"

I often responded, "On the day of the accident, I pleaded for space exploration to continue. It must, for all those who traveled that way before, for our nation to overcome its grief, and for our loved ones aboard *Challenger*."

The other *Challenger* families agreed with me. They too explained that the *Challenger* crew was on a pioneering mission to fly, explore, and teach. If their mission ceased with their deaths, then they had died in vain. We insisted that America's mission of exploration must continue. Rather than signaling the end, *Challenger* would serve as a transition chapter in America's story of pioneers on the space frontier.

The nation ached with a loss of its pioneer spirit. I was an ordinary person, a teacher myself, but I saw how the space community and the nation were hurting. Sometimes acceptance is too difficult and the cost too great. I knew something needed to happen to bring a phoenix out of the ashes. As always in times of adversity, I thought of my "ABC" guide to positive thinking. I was determined to use it to open the doors of opportunity to others and continue the important mission of the *Challenger* crew.

REDIRECTING OUR ENERGY

The healing was beginning. People throughout the nation approached our families to build memorials to the fallen astronauts. Enough requests, letters, and cards were mailed to us to fill a large room seven times over. The notes helped us with our attitude. Though it was extremely difficult, we learned to accept that our loved ones were gone and we must go on without them. We wanted to return the outpouring of love, especially to the children still waiting for lessons from space. Members from each *Challenger* astronaut's family came together in my home, sitting around the coffee table in my living room, to decide how to respond. Although most of the families lived in Houston, others came from California, New Hampshire, and Maryland to meet, talk, and plan.

I hired a secretary to help us get organized. The families selected me to be their chairperson. Together, we created a foundation, incorporating it as a non-profit organization for education. We called it a living memorial with no walls, no bricks, and no mortar.

The world knew that seven *Challenger* astronauts died, but those people were more than astronauts to us. They were our families and friends. The world knew how they died; we wanted the world to know how they lived and why they were willing to risk their lives. We named our foundation the Challenger Center for Space Science Education. On mighty wings, it would rise out of the ashes of tragedy to create a loving and living tribute to the *Challenger* Seven and to all people who still believe in dreaming, reaching for stars, and nurturing the pioneer spirit.

As we worked to flesh out our plan, we met several obstacles. Somewhat organized, we set out to induct leaders into our group and seek financial support. It wasn't easy, and I was often tempted to give up. Not everyone believed in our mission. To some we were simply widowed spouses who were reminders of a terrible space accident. At first, we went to speak with the public affairs team at NASA Johnson Space Center (JSC), then we called those at Kennedy Space Center. We thought either of those locations would be ideal for a child-oriented simulation experience.

However, I didn't understand how to work with big government entities, how to create a Memorandum of Understanding (MOU), or why they wouldn't want to help us do something so important and creative. We believed strongly in our mission, but they didn't. In fact, because the media reported frequently on our activities, one of the PR individuals said, "If we go along with your plan, it will be like the tail wagging the dog."

Perhaps they were right, but we still believed in our idea because it was the right thing to do, and we felt it could inspire many to believe once again in the potential of space exploration, especially a nation of schoolchildren. I had prayed for divine guidance, but now I prayed in earnest for a miracle. It was exhausting work to meet naysayers at every turn. But we were stubborn . . . or at least persistent.

SEEKING HELP FROM OTHERS

We had courage to keep going and committed ourselves to the idea. Along with clouds of frustration came unexpected glimmers of hope like my former high school student, Richard Garriott. He had become a young entrepreneur in the computer game business. I asked him if we could create an opportunity for students to work together in a game-like scenario to solve problems on a simulated space flight. Could each student work at a station, depending on each other, to complete a mission aboard a mock-up of a space station and mission control? In a youthful, can-do reply, Richard said, "It hasn't been done, but I don't know why not."

Additionally, our artist friend Bob McCall and his wife, Louise, helped us sketch a design for an interactive and high-tech educational center. With Richard's creative ideas and Bob's artistic talent and imagination, we created the concept to deliver the dream that we thought would help the nation continue the *Challenger* mission.

I traveled to California to meet with officials at Rockwell, the company that built the orbiter. They were generous with their time and loaned us an executive to help. We visited the set of the television show *Star Trek* to see their ship, the *Enterprise*, for ideas. Bob McCall introduced us to story boarding with the Disney Imagineers, who helped guide us in the development of a space-themed scenario. I traveled to visit leaders at Martin Marietta, the airplane company in Maryland, who provided much-needed seed money. We organized ourselves even better, established our mission statement, hired a small staff, and collected more funds.

Then I placed an important call to Vice President Bush. I remembered the slip of paper he had given us after the accident that offered his home phone number and the message, "Call if you need us." Months later we accepted his offer, not for us, but for our dream to see the *Challenger* mission completed.

When the vice president returned my call the next day, he said he liked our idea and invited me to come to DC for a meeting. I was overjoyed to hear his personal encouragement and support. One thing he said, though, made me curious, and I asked Rich why Mr. Bush would have commented that "quite a character" had answered our phone.

Blushing, Rich admitted somewhat shyly, "Honestly Mom, I didn't answer the phone *Scobee's Bar and Grill.*"

I grinned, having told him for years that his joke would back-fire some day.

"But," he continued, "when he said he was the vice president calling for Mrs. Scobee, I wanted to know . . . 'Well, the vice president of what?' But he laughed when he said, 'Of the United States.'"

We met on a glorious early spring morning in Washington, DC, sitting in comfortable chairs in his office. Vice President Bush specifically wanted to know how the children were doing since the loss. After a few moments of compassionate conversation, he asked, "June, how are your plans for the Challenger Center coming along?"

I explained our difficulties, particularly my inexperience. He got me to think about how I motivated students in my classroom. "I create a vision for my students of what we want to accomplish for the day or the unit of study," I explained, "and then I try to inspire them with the importance of the lesson."

Bush suggested that I do the same as chairperson of my board. He also advised me to use the media to my advantage rather than shrinking away from them. "They will help you get your message out," he encouraged. "Most of them are good people, and they'll be good to you."

Later, Vice President and Mrs. Bush hosted a reception at their home in the Naval Observatory to show their support for our ideas. This meeting helped raise our confidence.

Because Mr. and Mrs. Bush led the way with their early supportive response, others followed. Upon Mr. Bush's suggestion, we expanded our board of directors beyond the Houston group that included former director of JSC Gerry Griffin, astronaut Kathy Sullivan, George Nield from the Office of Advanced Manned Vehicles, aerobatic instructor pilot Debbie Rhiene, and University of Houston chancellor, Tom Stauffer.

NATIONAL GROWTH OF THE CHALLENGER CENTER

We hired a president for the Challenger Center, Jim Rosebush, whom we met when we were in Washington, DC. A strong advocate, he organized us even further beyond our non-profit status as a 501-C3 organization. He traveled with us to New York to help with fundraising activities and speaking engagements on television morning talk shows and the evening news. We set up our office in a small apartment building in DC, then at a King Street address in Alexandria, Virginia. We continued to search for a location for the big center we planned.

On one of my trips to DC, I met Robert McAdams, secretary for the Smithsonian, who advised that the Challenger Learning Centers needed to be in multiple places, not in only one building like the Air and Space Museum. He agreed to serve on our board of directors.

Vice President Bush took a giant leap of faith to help us. A firm believer in our mission, he served as our honorary chairman, helped us with advice, made phone calls, and wrote letters on our behalf, even giving us a personal check of financial support. Together, we asked others to support us or volunteer to join our team on the quest to overcome a great national tragedy.

Those who volunteered to help told me they remembered where they were and what they were doing when they heard the awful news about *Challenger* in January 1986. They described the streams of white smoke and fragments of the shuttle that filled the cold blue sky. Now they wanted to help by making a difference.

After these volunteers accepted the initial challenge and joined us in leading the way, others followed. Time and dollars have translated into educational programs that encourage our youth to reach for the stars and work hard to see their dreams come true. Because so many people cared, we were confident that the *Challenger* crew's mission would continue. The volunteers stood beside us in our grief now stood with us as partners on the frontier of education. Of course, we still had challenges to accept and lessons to learn as we tried to accomplish our costly, innovative goal.

ANNOUNCING THE CHALLENGER CENTER

On September 23, 1986, at an elementary school in Washington, DC, we made a public announcement to students and news media about our intentions to create the Challenger Center. The news fanned out across the United States and around the world.

Headlines the next day read, "Shuttle Families Plan Space Hub"; "Challenger Families Planning Space Center"; "Challenger Families Accept the Challenge"; "A New Quest: The Challenger Families Launch Educational Project in Memory of Astronauts." The news traveled over the wire for newspaper services and for television and radio audiences. People in the news media who had invaded our privacy to tell the tragic story became our friends, retelling our positive new adventure. Many who heard the news sent us notes of congratulations for creating a living tribute to our loved ones. They commended me for my courage to speak publicly before dozens of reporters and television cameras. Some were surprised by our persistence and resolve to see the dream become a reality.

Our plans were made public, but our prayers were private. We asked God for courage and guidance to know what was best for the *Challenger* Seven, our families, and the children around the world who were still waiting for their lessons from space. We certainly traveled challenging roads to create the Challenger Center and its programs. However, we knew the greatest risk was to take no risk. Our enthusiastic staff and my stubborn will were not swayed. Sheer determination and tenacity helped us create a worthy tribute to the *Challenger*. I continued to be hopeful that with our positive attitude, a strong belief in our mission, and our commitment to help dreams come true, we would find a silver lining hiding behind the dark clouds we had endured.

Letters like this one from Vice President Bush (December 1986) greatly encouraged us:

Dear June,

The loss of the *Challenger* was a loss for the entire country, but I am watching you and the families of the other crew members do a marvelous thing: rebuild the mission of flight 51-L and carry on with its work. The Center's memorial in honor of the *Challenger* crew members is an inspiration to the children, teachers, and future leaders of this nation. This useful knowledge will have a strong bearing on the future success of the national manned space flight program.

I'm sure you know you have my full support and personal best wishes in this endeavor.

Sincerely,
George Bush

FACING THE HOLIDAYS AND THE ONE-YEAR ANNIVERSARY

The Christmas holidays were difficult for all the *Challenger* families, just as they are for anyone who has lost a loved one. I phoned all the spouses to ask how they were getting along.

Some who had young children wanted to keep the same traditions as in the past, while others sought an escape from the reminders of their loneliness. Still others wanted to experience a true worship service celebrating Christ's birthday and God's mighty love. We were all different in our needs, but we each shared in our common loss and continued to support each other.

Continuing to organize our plans with confidence as we dove into the New Year, we decided to respond to the many media requests with personal heartfelt words. The following letter, written one year after the accident, represents our philosophy and reasoning behind the Challenger Center.

January 28, 1987
Letter to America from the families of the *Challenger* crew

One year ago, we shared a terrible loss with you. The *Challenger* crew were our husbands, wife, brothers, sisters, mother, fathers, daughters, and sons—the fundamental, irreplaceable people in the fabric of our lives. At the same time, they—their mission, their quest, their essence—were an intrinsic part of national life too, part of the great extended family known simply as "Americans." They were our pioneers. Together we mourned them and the shortness of their lives. But, in their short time, they contributed and left memories.

They were not people who cherished the soft and easy life, but people who worked hard to extend the reach of the human race no matter what the sacrifice. They risked their lives, not for the sake of aimless adventure, but for the nation that gave them opportunity, and for the space frontier which was an extension of its spirit. They were scientists, engineers, and teachers, guiding us to space. *Challenger's* mission—to give ordinary

Americans access to space, to push scientific discovery forward—was a culmination of their work, a fulfillment of their hopes, and an expression of their essential being.

Since their loss, we have been troubled by the incompleteness of their mission. Lessons were left untaught; scientific and engineering problems were left unsolved. Perhaps saddest of all is the idea that America's children must once again put their dreams and their excitement about the future "on hold." This is too great a loss, one we cannot accept.

We wish to carry on *Challenger's* mission by creating a network of space learning centers all over the United States called, cumulatively, the Challenger Center. We envision places where children, teachers, and citizens alike can touch the future. We see them manipulating equipment, conducting scientific experiments, solving problems, working together—immersing themselves in space-like surroundings and growing accustomed to space technology. As a team, they can practice the precise gestures and the rigorous procedures that will be required of them on the space frontier. They can embrace the vision and grasp the potential of space too.

Though it will take time and money to build, the Challenger Center is our idea of a fitting tribute, a celebration of our loved ones' lives, and a triumph over their loss. We hope that by making space-like experiences accessible to all people, especially children, we can prepare them for the day when they will take their own place among the stars.

If they were alive and could speak to all Americans, we believe the *Challenger* crew would say this: Do not fear risk. All exploration, all growth is a calculated risk. Without frontiers, civilizations stagnate. Without challenge, people cannot reach their highest selves. Only if we accept our problems as challenges can today's dreams become tomorrow's realities. Only if we're willing to walk over the edge can we become winners.

The team grew after we announced our mission. We received contributions from children who sent their nickels, dimes, and pennies (one sent his tooth fairy money); the elderly who still believed in the great American dream; and leaders in government, business, and education. To build the best, we needed the help of the brightest and most creative people available to help us. Still others called us to offer their support. Disney invited our families to a special day of recognition, a parade, and a major celebration at Disneyland. Lee Greenwood volunteered to serve as a board member and to sing his popular song, "God Bless the USA," at our first awards dinner. Our circle of friends grew until the ripples reached across America, around the world, and, most important, into the classrooms of students and teachers still waiting for their lessons from space.

President Reagan offered support in the following letter he sent to me and through his proclamation for National Challenger Center Day.[1]

Dear June,

    Ever since that tragic day last January, children everywhere have been waiting for their lessons from space. I warmly congratulate you on your tireless devotion to continuing *Challenger*'s mission and to providing these youngsters with the lessons that seemed lost.

    With the Challenger Center there will be a brighter future, full of hope for all our children to explore and to learn about the frontier of space.

    God bless you.

Sincerely,
Ronald Reagan

FIRST NATIONAL MEETING OF VOLUNTEERS

Kathy Sullivan, astronaut and friend, helped us convene our first national meeting in Tucson, Arizona, in spring 1987. We called on the scientific community, leading educators, computer experts, astronauts, mission specialists, and flight directors to help us design the educational program. Others joined us too: corporations, foundations, the media, and individuals who could help us with financial support.

To overcome not only the grief and sorrow for our loved ones' unfulfilled dreams, but also for the nation's loss, this massive team of volunteers helped us create more than a memorial. Together, we created Challenger Learning Centers, where students and their teachers can climb aboard a child-sized space station and fly a simulated space mission into the future, applying math and science skills while working together as a team to complete a successful mission.

Kathy led the group to consider building the center at science centers and museums, while I preferred that the centers be located on college campuses. Even with our different visions, we had a tremendously successful meeting and ended with a design we all approved.

Later, we met in Houston with museum leaders from across the nation to identify the director who could use our design, Bob McCall's painting, and an architectural rendering by a firm in New York City to assemble the first Challenger Learning Center. For weeks afterward, I contacted each director to discern who was most prepared to work with our staff to transform sketches and words into an interactive space simulation experience.

BUILDING THE PROTOTYPE

Carolyn Sumners of the Houston Museum of Natural Science was chosen as the creative leader to help a team assemble the collection of computers, cabinetry, and futuristic space parts that would become part of the first Challenger Learning Center. Immediately, she contacted a group of education leaders, museum developers, our staff and board members, and Richard Garriott. Together we finalized the design and, with the help of Vice President Bush, secured funding. Mr. Bush wrote letters on our behalf and recommended foundations that might be supportive. (The William Stamps Farish Fund responded with enthusiasm to help us fund the prototype Challenger Learning Center.) Carolyn also involved employees of Apple computers. I visited regularly and joyfully, watching our dream come true.

By 1988, the first Challenger Learning Center prototype was built in Houston ("Space City") with the lively local leadership of Carolyn Sumners and our dedicated and talented staff led by new science center leader David Winstead. Astronauts, NASA mission controllers, educators, and museum designers also joined the team.

The day we opened the doors to the public and media to launch our first center, I stepped through the air lock into a world of suspended disbelief. Better than my imagination, the center resembled a futuristic space station grander than the Hollywood set for shows like *Star Trek*, including a perfect child-sized replica of NASA Mission Control. We celebrated with grand fanfare.

A television reporter asked me, "What do you think the *Challenger* crew would think of this marvel?"

I was delighted to say, "I think Christa would make it one of her favorite field trips because it represents so much her philosophy of hands-on learning, and if this were in the neighborhood where Dick Scobee grew up, I imagine he would show up at the door regularly to 'fly' on a mission to the moon or Mars or rendezvous with a comet. I think they would be very happy. I don't think they would be surprised that we were able to continue the education mission for them."

MOVE TO WASHINGTON, DC

It grew increasingly difficult for me to meet with our staff in DC and at our center in Houston. I decided I needed an address at both locations. Another prayer was answered when Rose Narva, manager of the Jefferson Hotel, called to say we could meet there and offered me a permanent room at the cost of upkeep. Rose and her husband, Bill, a Navy admiral and the attending physician for Congress, became my dear and trusted friends.

Soon the centers grew in number, slowly at first, then at exponential rates: east to Maryland; south to Florida; north to Ohio, Connecticut, and New York; and then west to Hawaii, California, and Washington. Across our large wall map of the USA, we placed Challenger Center logos to represent the many locations.

Students throughout the United States travel to Challenger Learning Centers for a different kind of field trip outside the boundaries of our planet. Teachers climb aboard the space station with their students and fly a simulated space mission into the future, where they are challenged to apply skills in math and science learned in their classrooms. In this way, Challenger Learning Centers become a bridge between students and the hope and promise of a brighter future.

Our students are transported to a new world where they discover the importance of intense teamwork (in fact, many corporations send their employees to Challenger Learning Centers to enhance cooperation in the workplace). The children work together as a team of engineers, physicians, scientists, navigators, and communicators to chart flight patterns, assemble an electronic probe, solve oxygen and water shortage problems, monitor the health of the crew, and solve problems in a high-tech environment in a true-to-life experience. In the process, they increase critical thinking abilities and learn about the fields of science and math. Most of all, they learn about themselves and how to work with others in the world around them.

NEW YORK CHALLENGER LEARNING CENTER

During the grand opening of the Challenger Learning Center in Rochester, New York in February 1992, a teacher entered the space station simulator with her students. When she saw the reporters, photographers, and television cameras, she asked a boy named Jerome, one of her more animated students, to wait outside. The flight director and I encouraged her to let Jerome join his classmates. She hesitated, insisting that Jerome, who had been "in the courtroom more than in the classroom," would be an embarrassment. We explained that there was something at Challenger Center for all students, including Jerome. Reluctantly, she admitted him, but placed him in the enclosed "probe assembly" room.

Jerome wasn't much of a reader, but he was good with his hands. He experimented with the mechanical and electronic equipment until, at the urging of his teammates, he finally assembled the probe. When the mission was completed and the students congratulated each other for their success, the media zoomed in on Jerome, who was in his customary state of animation. When one reporter asked him about his duties during the mission, Jerome stepped forward, used the language of the engineer, and took all the credit for the successful mission. His classmates cheered!

Later, I learned that Jerome was still a challenge in the classroom, but for different reasons. His teachers told me he was eligible for a scholarship to college, and he wanted to know the name of the best school of engineering. For me, he is living proof that the Challenger Center experience is like no other to encourage our youth to reach for the stars and work together.

More than a classroom of computers and high-tech equipment, Challenger Center is a system that brings community leaders together to benefit their local schools by making learning exciting and providing innovative tools for teaching. Individuals, foundations, corporations, and companies work in a team effort to create the experience for their schools, science centers, or colleges and universities. It is a fantastic way to take education into the twenty-first-century space age of communication and technology.

THE CONTINUING *CHALLENGER* MISSION

Franklin D. Roosevelt once said, "We cannot always build the future for our youth, but we can build our youth for the future." That's what we all accomplished when we created Challenger Center. We made a difference. We touched the future not only for the crew of 51-L who set out to explore, inspire, learn, and teach, but also for generations of children who will work hard to see their dreams come true.

The *Challenger* mission continues each day when children arrive at the doors of a Challenger Learning Center. I see it in their broad smiles. I hear it in their serious commands, scientific reports, and cheers of jubilation. I sense it when I watch them engage math and science and when their teachers tell me delightful stories of increased self-esteem for underprivileged children or of complex problems solved through a team approach.

Most of all, I know the mission continues when I glance to the heavens. Sometimes I see seven stars twinkle and know seven great people look down on us and smile. When I hear thunderous bursts from the clouds above, I wonder if it isn't applause for the thousands of friends, sponsors, and supporters whose gifts of love have made the difference. If I use my imagination, I can hear the commander's steady voice communicating to mission control: "Throttle up. All systems go." I can hear Christa's enthusiastic voice rippling through space and time, saying, "Our lesson for today is . . . ."

Certainly, clouds of despair often threaten our world, frighten our families, or shatter our hearts. These clouds may come in the form of great illness or suffering, death or divorce, fear or anger. They may reach across a country and around the world. They may be ominous, menacing, and threatening. But clouds are not forever. Just beyond them is a bright light radiating with the warmth of God's love—a silver lining.

Note

1. The text of President Reagan's formal proclamation can be found in the appendix.

Chapter 17

FROM SCARS TO STARS

LESSONS LEARNED

Through our experiences both with the loss of loved ones and the challenge of beginning a major venture, we learned lessons on forgiveness, overcoming adversity, skirting obstacles, solving problems, seeking God's help, forfeiting control to God, and welcoming new life. How to overcome a tragedy, adversity, or a crisis isn't as simple as how to turn a not-so-good day into one to celebrate, but they are based on the same principle of *Attitude*. It isn't simple, but I learned that to make a positive out of a negative, you must first turn the energy of grief and sadness into the energy of creativity.

Attitude was very important. We learned to accept our problems as challenges, and through all the tribulation, we still *believed* in our mission and the power of prayer. We made a *commitment* to see our dreams for the *Challenger* crew fulfilled, but it took courage, especially the staying power of perseverance. Our reward was in seeing how much the students enjoyed their simulated missions at the Challenger Learning Centers.

We learned that life is a series of lessons and challenges that help us to grow. One of these lessons is how to overcome a crisis. The Chinese have a word for crisis that explains the process well. It consists of two pictured characters: one means trouble and the other means opportunity. When we meet a crisis, tragedy, or trouble, we can strive to turn the energy used for anger, fear, or grief into energy to solve problems and create an opportunity that brings triumph out of tragedy.

MY GOLDEN OPPORTUNITY

Our first Challenger Center president, Jim Rosebush, hired Rick Hutto to serve as vice president for development. A young man originally from Georgia, Rick had worked closely with President Carter's family and was familiar with Washington, DC. We worked well together, and I appreciated his willingness to dedicate himself to the *Challenger* mission. It was a joke, but he always greeted me as "Boss Lady."

One busy day as we traveled to a development meeting, Rick asked how I felt about attending a reception hosted by Guy Hunt, the governor of Alabama, in honor of "Dr. June Scobee Day." Rick wondered if I wanted to invite anyone specific to the event. I immediately thought of my former classmates from the school at Odenville, Alabama. In particular, I hoped to see the ones who had doubted me as a young teenager. Though I had long ago forgiven them, I hoped to come to terms with a painful past and reorient myself to a future that could break the tension between memory and hope.

On my special day, the governor's proclamation was read to a large group of people; many of them came from my high school. It was a wonderful reunion with students I had barely known—but with whom I felt a great kinship. I was delighted to learn about their individual exploits and accomplishments, but I was surprised to learn that only two of us from my class went on to college, and that I was the only one who became a schoolteacher.

After the event, reporters asked to meet some of my former teenage friends to learn what I was like as a young girl. I heard one of my former classmates answer, "Oh, yes, she was nice back then just like she is now." That day, I learned that we can't leave behind remnants of our past. Instead, we can embrace them as a part of us that helps us grow.

ALWAYS LEARNING LESSONS

Since my husband's death, I had discovered, as English author Matthew Arnold expressed it, "to see life steadily and to see it whole." As I moved forward with this philosophy, I still faced tough lessons. One of the toughest was learning how to gain self-confidence without being overconfident. After Vice President Bush encouraged *Challenger* families to work with the media as a positive way to spread our message, several of us attended work sessions with organizations that provide public relations training. One group helped us address personal questions that had the potential to blindside us. This group showed us how to focus on our message even when faced with difficult issues.

After our instruction provided by the public relations firms, I felt certain that I could speak with confidence to any reporter and remain undeterred from my intended message. Before I ever spoke on national television, a local news reporter from the *Houston Chronicle* asked to meet with me. Though others warned me that he was manipulative, I assured them that my PR training had prepared me

The reporter met me at our Clear Lake City office on one of the rainiest days of a week-long torrent of rain storms. My assistant invited him into the small office and asked him to be seated across the conference table from me, and for at least an hour we talked about building the Challenger Center. I stayed on message seriously and business like, though I smiled periodically at the kindness of the reporter and his questions.

When the interview ended, I walked the reporter to the door to say good-bye. As we stood at the door, a downpour of rain decreased visibility so much that I encouraged him to wait safely in the office.

"Is all this rain causing you any problems?" he asked.

"Not really," I assured him, "but it will make it more difficult for me to get home later so I can keep an appointment to meet with a roofer."

"You got a leak?" he wondered.

"Yes, from all this rain. I need him to repair the roof where I have a small leak in the ceiling."

With that, the reporter shook my hand, thanked me for the interview, and left my office.

The following morning, I opened the newspaper at the table, sipping a cup of hot coffee. Startled beyond belief, I read a large headline about myself: "Challenger Widow Repairs Own Roof." The article included nothing from my Challenger Center interview with the reporter, only a weeping widow story. Furious, I tossed the paper aside and cried. My overconfidence, almost arrogance, had led me to ignore warnings about this man and ended up wounding my pride. "I'm so stupid!" I shouted aloud to nobody. "Who am I anyway? Just a dim-witted girl from Odenville!"

Once I collected myself, I realized this was a major lesson I learned the hard way. There is definitely a fine line between self-confidence and brazen overconfidence. I could only blame my careless trust and foolish pride. I figured that at least the article appeared only in our local paper.

However, within hours as the time zone moved west, friends and family began calling me as they read their morning papers. The AP wire news system had picked up the story, which now appeared in newspapers all over the country.

The most concerned person who called was Dick's father. "Junie, I'm on my way to Houston to help you with your roof," he said. "I can't let my girl have problems. It looks bad that we aren't helping out our daughter."

"No, Dad." I assured him that the leak was minor and a roofer was hired. "Everything is okay. *I'm* the problem. The only damage is to my bruised ego."

BALANCING IDENTITIES

Dad laughed, reassured that all was well, but added, "Mother and I have lost one of our two sons, and we never had a daughter, so we wanted to let you know that we've decided to claim you as ours if that's okay."

I told him it was an honor for them to think of me that way. To my wounded heart, it sounded much better than "our son's widow." Six months after the accident, it was hard for me to view myself as a widow. I preferred to be the wife of the commander who died on *Challenger*.

In my estimation, I was still a wife, but in reality, I didn't know who I was. My new life existed in unfamiliar territory. Slowly, I learned lessons and gained strength as I began to accept my unwanted identity. I no longer absentmindedly set a place at the table for Dick, and I stopping getting emotional when I walked down the grocery aisle past Dick's favorite foods. But I missed him. I knew he was with me in spirit, but I wanted to hear his voice. I wanted to touch him and sleep cuddled up next to him. He could add comfort and help me solve problems, or at least make me laugh at myself and my foolish mistakes. My best friend and love was gone. Even now, as I write these words twenty-five years later, it is still hard to believe.

CHALLENGER CENTER SUPPORTERS

During my daily struggle to balance my identities and find the best fit, I continued to fulfill my passionate dream of expanding support for Challenger Learning Centers around the nation. As we worked toward building and expanding the Learning Centers, events sometimes put me in touch with celebrities and other powerful individuals.

Walter Cronkite

Dick and I had often watched Walter Cronkite on CBS. He was a trusted newsman who loved the space program and hailed from Houston. In spring 1987, before the first Challenger Learning Center was ever built, the Challenger Center staff, board of directors, and I decided I should visit Mr. Cronkite in New York City.

After setting up an appointment and traveling to the city, we met with several foundation and business leaders to discuss our goals, and then headed to Walter Cronkite's office. He met us at the door.

"Come in and look at some of my favorite NASA memorabilia," he encouraged.

Pictures given to him from famous astronauts hung on his walls, and space vehicle and rocket replicas adorned his desk and cabinets. I was intrigued by his museum of artifacts.

In my arms, I awkwardly carried a cardboard 3-D replica of a Challenger Learning Center simulator created by a New York design firm. Mr. Cronkite was curious and wanted to know more about our tribute to the *Challenger* "Teacher in Space" mission.

"Where are you going to build it, and what are the children going to do at the center?" asked Mr. Cronkite.

I explained the simulation experiences.

He interrupted, "Are you going to have opportunities for students to be 'communicators' on board the spacecraft, and will a station be devoted to a student news reporter or journalist?"

"If we can get the whole thing funded, that will be at the top of our wish list," I assured him.

"What can I do to help?" he asked.

As often happened, nearly every time I wasn't prepared or needed divine guidance, an idea popped in my mind. I invited Mr. Cronkite to emcee our first big gala dinner, where we planned to thank many people for their support.

To my delight, he agreed.

Later, Mr. Cronkite accepted my invitation to serve on our national Challenger Center Board of Advisors. He personally shared our enthusiasm to fulfill the education mission.

Oprah Winfrey

On the afternoon before the gala, Oprah Winfrey was in DC and invited the *Challenger* spouses to appear on her show, where we were welcomed by a cheering live audience. All spouses were present to be interviewed except Steve McAuliffe, so Grace Corrigan, Christa's mother, represented her daughter. Judy's brother, Chuck, represented her. Oprah's staff was professional and handled our group of seven kindly. Even with her thoughtful questions, Oprah had a difficult time steering us away from our purpose of spreading word about the Challenger Center.

During commercial breaks, she asked us to speak about the *Challenger* explosion, but we explained that we were there to share something else. "Please let us talk about our tribute," I pleaded. "Everyone knows how they died. We want them to know how they lived and about our Challenger Center."

During the second commercial break, even the audience encouraged Oprah to let us talk about our tribute. When the cameras rolled again, Oprah asked us to look at the television monitor and talk about what we saw. I was astonished to see the film footage of the space shuttle explosion. Once again, we had to watch the as the shuttle carrying our loved ones exploded and broke into a million pieces.

Instead of weeping before the cameras, we grew despondent and solemn. Frustrated, Oprah finally gave in. "Okay! Tell us about your Challenger Center."

Afterward, none of us were pleased with the interview. I felt dejected, wondering what we should have done. As we left the studio, I stopped to ask Oprah, "Would you like to join us at a formal dinner this evening?" She seemed pleased to be invited.

The Gala Dinner

That evening at the dinner, our Challenger Center president, Jim Rosebush, asked Oprah Winfrey if she would like to help give out the *Challenger 7* awards to the seven people the *Challenger* families (founding directors) had selected for their contributions to the mission of the Challenger Center. She graciously agreed.

During dinner, Annie and John Glenn sat at my table, and Walter Cronkite sat with my daughter, Kathie, who had been a journalism major. At the gala, Vice President Bush and Senator John Glenn spoke, and Oprah presented the awards. Lee Greenwood invited the *Challenger* families to join him on stage to sing "God Bless the USA." As we stood next to each other singing, I looked out to the audience and the bright lights overhead and silently said a prayer of thanksgiving for those who joined our efforts and came to our dinner. They came from California, Texas, Florida, and many other places across our country to help us pay tribute to our loved ones. We were humbled and ecstatic.

MAKING A MOVE

These celebrity encounters and the gala were among many events that helped further our mission. Along the way, I gained confidence in navigating the airlines and making close connections at airports. However, traveling between two cities became difficult, expensive, and tiresome. The Challenger Center office was located in Washington DC, and my home was in Houston. When I saw other astronaut families in Houston, I felt as though I was a constant reminder to them of the *Challenger* accident. For me, seeing them was a reminder that I no longer had a family to come home to after work. Our son, Rich, had graduated from the Air Force Academy and was excited about flying F-16s for the Air Force. Our daughter was married and working as a public relations director for Texas A&M University. Even our dog and cat, our pets for many years, had both died.

Though the Narvas had generously donated a room at the Jefferson Hotel in DC, after two years as their guest I finally chose to sell my house in Houston and purchase a home in or near Washington. After searching, I made the decision to buy a small townhouse under construction in Georgetown. In the meantime, I stayed in my room at the Jefferson.

As always, Rose and her kind and professional staff were generous with conversations when I returned from the office to relax in the sitting room. I also met interesting and lovely people at the hotel. Rose or her husband, Bill, often asked me to meet a special guest traveling from fascinating cities around the world. Sometimes they introduced me to dignitaries or to news correspondents. One such person was Larry King of CNN. After our initial meeting, he invited me for an interview on his show, which I cautiously accepted.

THE LARRY KING INTERVIEW

Within days, a driver provided by CNN took me to the studio to meet with Larry King. Uneasy, I looked around while Mr. King studied his notes. Just after I was asked to sit across the table from him, the questions began. I was fairly comfortable as we talked. Then he arrived at the final question.

"Do you think you will ever remarry?"

Instantly, I felt my casual façade turn to one of chiseled stone. Tension rose in my voice as I answered him with grit. "Dick Scobee loved me enough to last a lifetime. No, I will not marry again."

Just as quickly, my anxious edge relaxed, and I was myself again. We ended the program on a high note of hope and expectations.

Back in my hotel room, I climbed in bed and thought about the interview question that rattled my self-concept. Why was I so insulted? Larry King was not insensitive. He asked a fair question that caught me off guard, but it told me I lived with the perception that I was still married to Dick Scobee. Though I had heard Dick's encouragement to press on in my dreams, I had not accepted the reality that Dick was gone from my life . . . forever. I sobbed into my pillow until I slept.

SCARS TO STARS

The following morning, I dressed in casual slacks, slipped on comfortable shoes and a warm coat, and left the hotel in the city to go to "Old Town" Alexandria, Virginia, where our Challenger Center office is located. Instead of driving toward the office, though, I turned toward the Potomac River, parked my car at the water's edge, and walked along a path.

I kicked rocks along the way and felt sorry for myself. I dredged up every scar from my childhood and those of the recent years. "Who are you, June?" I asked myself. "You are no longer a wife; you are a widow." It seemed that, while I worked so hard to build Challenger Center and complete the lessons from space, I'd missed out on a few lessons I needed to learn for myself. I wondered what was next.

With God's patience, I began to learn that the lessons were not there to defeat me, but to help me grow beyond old habits of holding on to loss, seeking perfection, and keeping up appearances. I also knew I needed a friend. I wanted to experience the joy of an intimate conversation that would nourish my soul and fuel my senses. Through these necessary parts of a joyful life and with God's help, I knew I could learn to move past the scars and reach for the stars. All of these are moments to treasure when the burdens of yesterday give way to the joys of today. Life—more to live, joys to discover that reach the heights equal to but opposite from the depths of despair. That's when I made the decision, "I'm forgetting the scars, and if I reach tippy toe high enough, I'll see the stars." I've been out there with the clouds of life, acquainted myself with their many faces, from the calm to the turbulent. I've known their beauty as well as their wrath.

God saw me through the depths of grief and through the struggle of tribulations. Today, on my walk, on my escape to find solitude in nature, God helped me to solve problems and weather the storms of anger, guilt, and self-pity—the storms I battle openly and those I fight inwardly to test my faith, pride, and humility. Greatest of all, I learned lessons about love, the beautiful power of love that can cause a magnificent phoenix to rise up out of the ashes of a tragedy.

When we reach for the stars; what are we expressing? Is it persistence to keep going? Or a more determined reach that overcomes obstacles like perseverance? Is it to triumph when all about you is crumbling, or is it simply the passion of fulfilling a dream?

Chapter 18

HARMONY OF LIGHT AND LOVE

GETTING BEYOND THE GRIEF

By the second year after their father's death, my children had learned to reach for the stars. Rich completed his final semester of college, graduated from the Air Force Academy, went on to pilot training, and then married his high school sweetheart. Kathie carved out a beautiful career as a public relations director for colleges and created a lovely home for her family. It was time for me to move forward too.

During the weekdays, I lost, even buried, myself in the work of first the university, then Challenger Center. Sundays, though, always had been reserved for family and were especially lonely for me those first two years. I didn't know how to change that tradition and had not made the effort to make new friends.

After I moved from Houston to Washington DC, my children and other family members called regularly. Sometimes, I heard from friends: Nancy who cared and made me laugh, Pat who knew how to picnic even on a rainy day, Elaine who made arrangements with a florist to send me a fresh bouquet of flowers for several weeks, and Diane and Kirby who encouraged me. Kathy was thoughtful and fun to talk with. So many others wrote meaningful notes that made me smile. But they were all miles away in Texas, California, England, and Germany. I missed them, our conversations, and having dinner with someone who cared.

During the second anniversary of the loss of *Challenger*, several reporters interviewed me for television, news, or magazine stories. One was Jane Mahaffey, a young widow herself and a talented writer. Jane invited me to an Easter sunrise service to be held at Arlington National Cemetery where Dick was buried. Having no other plans, I accepted without hesitation. Jane mentioned that other widows and widowers would also attend, including one man in particular who had lost his wife only the month before. We all needed the support of friends who understood.

That Easter morning, in my hotel residence, the alarm clock buzzed with loud indignity. Groping in the dark, I switched off the alarm and turned over to sleep again, tugging at the cover for warmth. I'd already decided it was too dark to be alone on the streets and too cold to sit shivering through an outdoor sunrise service. I would explain my absence, and Jane would understand. I dozed until a bolt of guilt flashed through my conscience and brought me quickly to my feet. Racing to get dressed and out the door, I thanked God for his saving grace and for the new friend whom I didn't want to disappoint.

EASTER AT ARLINGTON NATIONAL CEMETERY

When I arrived, Jane was waiting to meet me. It was still dark when we met in the amphitheater at Arlington National Cemetery and joined other widowed people. We greeted each other solemnly and took our seats, watching quietly and reverently as the morning light began creeping over the horizon.

The chaplain opened the service. "Easter sunrise observance marks the dawn of Easter, Christ's promise of new life for all people," he told the congregation. "It's the most important holy day of the Christian religion."

The dark sky gave way to light. The sun, a rosy orange ball, first peeked through the Greek architecture of the amphitheater, then turned to a yellow glow, climbed a sky of scattered clouds, and spread its rays gently over us where we sat worshiping together.

As I listened to the message, I was encouraged. I thanked God for the challenge of yesterday, the beauty of the day, and the hope of a new tomorrow. I was ready to let go of my vulnerability and reach out to the people weeping softly beside me, but the energy I used for my own sorrow sapped my strength to help the others.

With a strong, determined voice, the chaplain said, "To have new life, new friends, new experiences, we must let go of the pain and close the door to the past. We must learn to trust God to lead us safely through to the new day." I scribbled his words on the back of my bulletin because they gave me courage to reach out to the others, to smile, and to nod my understanding to them.

After the service, I walked across the street from the amphitheater to Dick's gravesite and wished him a happy Easter. I returned to the others to find that another guest, General Don Rodgers, had invited us to join him for breakfast at the Ft. Meyer Officers' Club. About ten of us sat around a long table eating and getting acquainted. General Rodgers sat across from me and asked that I call him Don. He told me he worked at the Pentagon as the Director of Army Command Control, Communication, and Computers, but that he would soon transfer to Arizona.

We talked easily with each other. He had as many compassionate questions about Dick's death as I had about the sudden death of his wife, Faye. From his southern accent, I guessed he grew up in the South even before he told me he was from Cookeville, Tennessee, a town only about three or four hours from where I had lived in Alabama.

We ate the meal with pleasant conversation, but as we prepared to say good-bye, Jane suggested a walk along the Potomac River. A few of us agreed. As we tried to figure out the logistics of changing out of our Sunday clothes and meeting again, General Rodgers, decked out in his Army uniform with ribbons and three stars, calmly laid out the plan, the location, and the directions. For the first time, I really looked at him. He was tall and slender, with piercing hazel eyes and a gentle voice. In Don, I saw a man of admirable strength in spite of his recent loss.

Soon our little group reunited to walk along the river on a well-worn footpath. Unexpectedly the others decided to jog, leaving me walking alone with the general. We talked more openly with each other, sympathizing with each other's losses and our concerns for the future. We felt peaceful together on that beautiful blue-sky day.

I learned that Don had lost Faye to a sudden heart attack as she drove home on a crowded Washington freeway. A fellow traveler pulled off the highway and parked behind her car to help, calling for an ambulance when he saw her slumped across the steering wheel. A police officer phoned Don's office at the Pentagon to tell him the news, and another came to rush Don to his dying wife in the emergency room of an Alexandria hospital.

Seeing his wife's lifeless body attached to tubes, a breathing apparatus, and electronic devices was traumatic for Don. Her doctor offered no hope or encouragement. Don notified their only son, Eric, who flew from Dallas to Washington to his mother's bedside. In only a few hours, Eric was there embracing his father and they held his mother until all life left her body.

Walking together by the river, Don and I appreciated each other more, knowing the sorrow that filled our hearts. Don's loss was not public like mine, but it was just as sorrowful for those who loved Faye. We agreed that at first we cry out in anguish for our dying loved ones, and then for ourselves—for the gaping wound the loss leaves. Our conversation was comforting.

We returned to the cars where the others waited and thanked each other for a full day of companionship. A few weeks later, Don would relocate to command Fort Huachuca near Tucson. "Thank you, June," he said to me before driving away. I never expected to see any of them again except for Jane. Even so, I was certain that day would influence the rest of my life.

TRANSITIONS

As the freshness of spring gave way to the heat of summer, work continued to fill my days. I traveled across the country, seeking support for Challenger Center and visiting my children. By August, Rich had an assignment to fly F-16 fighter jets at Ramstein Air Force Base in Germany. It was difficult to tell him good-bye.

Even with my renewed sense of purpose and direction, I often experienced moments of despair. To fight back tears of loneliness and the fear of isolation, I immersed myself in work even more, slipping into my old way of relying on my selfish need to control every situation and outcome. When a company created a television movie about the *Challenger* accident that none of the *Challenger* families approved, the stress of trying to prevent its release became so overwhelming that I needed medical attention.

With no neighbors to call and my children in other parts of the world, I asked Jane Mahaffey to help me. As we drove to the hospital, I prayed for guidance, for strength to let go, and for wisdom to overcome the obstacles in my path. Jane helped with her presence, listening and cheering me with jokes. She helped me create a humorous image of myself standing atop the world trying to save everybody else, when I was actually the one who needed saving. She invited me to talk to our chaplain at the Army post and to a psychologist. These individuals led me to understand that no one is perfect, that feigning great strength is actually a weakness, and that my need to hold on to the past (particularly to Dick) was a selfish act. For my health and well-being, I learned that I must first turn away from grief and loss and turn toward God and the gift of new life.

Guided by the light of love and laughter that healed and softened the jagged edges in my life, I soon had the strength and courage to begin letting go of the need to hold on to my past and control others, and to begin to create time and opportunities to make new friends.

After church the next Sunday, I drove to the cemetery where Dick was buried. I knelt at his grave and prayed for strength to accept my life as it was, to begin to move forward, and to turn control back over to God. I needed to hold on to the memories while letting go of the past. I needed to let Dick Scobee return to his star that would glow in the heavens forever.

A NEW BEGINNING

Several weeks later, Brian, Don Rodgers's military aide, called for him and told me the general would like to speak with me. When I answered that the general could call me himself, Brian's voice softened as he told me Don would be in DC for meetings at the Pentagon and wanted to see me. He had heard that the nice lady he met at Easter had been in the hospital.

Rather nonchalantly, I invited Don to dinner in my new home at Georgetown. When he stepped through my front door, he wore casual clothes and had shed his business-like exterior. He fidgeted while I cooked tenderloin steaks, potatoes, and asparagus, so I asked him to check a warning light that had been flashing in my car. After adding water to my radiator, he continued to tinker, repairing a loose cabinet door while we listened to my favorite radio station.

I carried the hot food to the table and allowed Don to seat me with a gentleman's flair. With a voracious appetite, Don ate his meal and part of mine as well. Only after eating the beef and rich sauces did he tell me he'd been on a low-cholesterol diet for some time. We laughed at his indulgence. After dinner, we sang along to a radio song, and then, with a sensitive smile, Don took my hand and twirled me around the kitchen. It was fun to laugh again. Later, as Don turned to go, he gave me his personal home phone number and asked if he could see me again.

"It would be wonderful," I said, "but no more high-cholesterol dinners for you!"

As chairperson of the Challenger Center, I had a tremendous amount of responsibility to work with our board of directors and small staff and to meet with leaders in education, business, and government. I frequently traveled to meetings and to grand openings of new Challenger Learning Centers and then always returned to an empty house. Fall arrived quickly in DC, filtering the thick city air and cooling the nights.

Unexpectedly, several weeks later Don Rodgers personally called to ask if he could see me for dinner when he flew into DC after visiting his parents in Tennessee. As I waited to pick him up at the airport, I wondered if I would even recognize him. After all, I'd only seen him twice. Finally, turning the corner and coming into view was a familiar face, but it wasn't the tall, distinguished general I remembered. Instead, this man wore a smile across an ashen complexion. Looking unusually thin in his slim jeans, he walked toward me. Sadness filled his eyes. He was alone, with no military aides or friends around him. Dressed in my pale pink suit, I reached out to hug him.

Standing there together, we must have looked ghostly: two lonely people, survivors whose paths had each ended abruptly, taken a turn on different journeys, and now had come together. We each needed a friend—someone who cared, someone who would call, someone who would understand. We had both married our high school sweethearts, so being rather naïve about dating, we simply grew to be good friends.

We spent a few hours talking after dinner and comparing our pasts. When I wanted to understand more about how the Air Force compared to the Army, Don created a flow chart on his napkin, explaining the organization and functions of the various branches of the Army, elaborating on the mission of the Signal Corps, his branch in the Army, and telling me about the people under his command at Ft. Huachuca in Arizona. Our conversation was relaxed and natural; I felt safe and happy.

Our friendship grew. We talked on the telephone like teenagers—getting acquainted and learning about each other's families, interests, Christian faith, joy in each other, and hopes for the future. As we healed, we learned to laugh more and to appreciate each other. Our paths had met and stretched out before us as one. I was forty-six years old, and Don was fifty-four. He and Faye had lived a beautiful life together for thirty-two years. Dick and I were married for more than twenty-six years. Could we possibly begin a new life together? We wondered, but not for long.

At a meeting I attended, I met another general and asked if he knew Don Rodgers. "General Rodgers is not like any other general," the man said. "He's serious about work, but no one has more fun in life."

I wanted to know about Don and Faye. The friend told me, "They were the ideal couple. They were high school sweethearts and totally in love and dedicated to each other. That Rodgers was a great family man."

THE SERIOUS QUESTION

I told Don that learning of his love for Faye made me love him more, and he asked, "Would you ever consider remarrying?"

The question did not disturb me as it had when Larry King asked it, so I thoughtfully responded, "Dick Scobee was my soul mate like Faye was yours. I will always love the man, especially the treasured memories of him as the father of my children. If I marry again, there has to be room for love to grow to allow me the memory of my love for him and my continued dedication to the education mission of Challenger Center."

Don liked what I said and voiced the same feelings. "Love does not die along with the person," he said. "I will treasure the memories of my wife, especially those of her as a mother of our son. Yes, I agree that there's room for love to grow."

The tables turned as my children took on the role of parents. Don, the perfect gentleman, asked my daughter if he could call on me for dates. Kathie giggled at the thought of her mother dating. For several months, Don visited me while he worked in DC. Then, for Thanksgiving, Don invited me to visit him at the guest quarters at Ft. Huachuca. I agreed, but I called Mom and Dad Scobee and asked what they thought. They were happy for me and encouraged the visit. I also called Louise and Bob McCall, who created the original sketch of Challenger Center and who lived in Phoenix, and asked if they would join us for dinner in Tucson.

It was delightful to see Don interacting with my friends. We laughed and shared stories and memories. As we left the restaurant, Louise gave me the thumbs up and said, "We both like him. Invite us to the wedding."

I blushed as the McCalls hugged me good-bye and Don escorted me to his sports car. We drove from Tucson to Ft. Huachuca to the military guesthouse on the post. The next morning I awoke to the sound of reveille and cannons and saw a magnificent view of a grand old-world military Army post with a grassy parade ground and a gazebo. Across the field, I saw a soldier raising a giant flag up the flagpole next to the smoking cannon.

Later that morning when Don called for me, we walked out to see a breathtaking range of mountains. We climbed to the top of one and looked down on quiet desert valleys. Deer walked along with us, undaunted by our presence, and eagles flew overhead. Don told me that we were so far south in Arizona that the Mexico border lay only a few miles below.

I felt a lovely kinship with everyone I met on the post. From Don's secretary Betty Olsen to his house aide Eddie Sneed to his family, they all admired and respected the general. He made everyone comfortable, often telling jokes about himself. Laughing brought out the sunshine in my life that had been hibernating in the clouds.

Returning home, I followed up with plans to speak at a conference in Vienna, Austria. Before I left the country and also while I was away, Don and I spoke often on the phone. It meant so much to know someone cared.

GREAT STEPS FORWARD

As Don and I took steps forward together, NASA celebrated a return to flight. Bill Readdy and Ken Reightler were in the first class of astronauts selected right after the *Challenger* accident. Prior to the launch, both worked on "Return to Flight" STS-26. This shuttle mission, the first since the loss of *Challenger*, was launched in September 1988. Later in the fall, their crew, led by Rick Hauck, was invited to participate in a parade at Disneyland along with the families of the *Challenger* crew. There we met Leonard Nimoy of *Star Trek* fame and other dignitaries. I was delighted to feel like a kid again and enjoy the amusement park rides with two dear grandchildren.

When I returned home from California, I received an invitation to a holiday reception at the home of Vice President and Barbara Bush. I looked forward to seeing my dear friends and congratulating George on his recent election as president of the United States.

In the meantime, Don asked my son, Rich, if he could marry me. Later, Rich told me he answered, "Well, that's up to my mom." Rick and I laughed and embraced each other's new joy and new beginnings.

My heart felt joyful as I watched the city decorate for Christmas. On the night of the Bush reception, I drove to the circle drive that leads to the front door. After turning over my car to a valet, I entered a house decorated with Christmas finery and brimming with aromas of good food. I spoke to a few people I recognized, then joined others waiting in the receiving line to greet the Bushes and have a photograph taken with them.

When it was my turn to step forward, I said, "Congratulations! I'm so happy you were elected president, and Mrs. Bush, you will be a wonderful First Lady for our nation." As I turned to step away, I whispered to them, "I have a surprise to tell you."

The vice president stopped me and said, "We already know. You are getting married."

I was never so surprised in my life. "How do you know that? What do you think?" I asked, grinning a broad and blushing smile.

Mr. Bush whispered, "I overheard my driver talking about the *Challenger* commander's widow and General Rodgers getting married. Our driver shared the story!"

Then Barbara added, smiling, "You know we must meet your general soon for our approval." We all laughed.

Within days, I flew to Ft. Huachuca to spend Christmas with Don. He wanted me to see how the Post Officers' Wives Club had volunteered to decorate his house for the holidays. Stepping into his home, I saw blinking lights, colorful ornaments, and garland stretched around a charming tree. Before he walked me to the guest quarters, he asked if I would join him for a glass of champagne. We toasted and sipped the sparkling drink, and Don seemed eager to have me sit with him in front of the tree. After I kneeled to look at the gifts, he asked if I would like to open one gift early.

I giggled, feeling like a kid with expectations of sugarplums dancing in my head. Don gestured to a small, gaily wrapped package balanced on a tree branch. I took the box in my hand and opened it to find a beautiful gold ring with a sapphire stone. He had purchased it while on business in Italy.

With the kindest expression and sincerest words, Don placed the ring on my finger and asked me to marry him. We promised we had enough love to go around for each other, for all our children, and for both Faye Rodgers and Dick Scobee. Don had already set the wedding date for June 3, 1989. We would marry in DC at the chapel near Arlington where Dick was buried and where Faye had her funeral service. After the wedding, we planned to have one reception in DC and another in Arizona.

A SPECIAL INVITATION

Shortly after the holidays, I received an invitation requesting my presence at the presidential inauguration for George H. W. Bush. Because I had a speaking engagement on the same date in Hamburg, Germany, I sadly sent my regrets.

Weeks later, when I returned from Europe, another invitation arrived, inviting Dr. June Scobee to the White House on the fourth of April for the president's first state dinner honoring the Egyptian president and his wife. Don had also received an invitation to the White House dinner, so we made plans to attend together. I teased, "You know you have to meet with their approval before you can marry me. That's why we are invited, so I can introduce you."

Eventually, the winter chill left the DC region, replaced by springtime warmth that burst upon the city with cherry blossom trees circling the reflection pool at the Jefferson Memorial. My townhouse was located not far from that marvelous display of color on the National Mall, so I enjoyed the beauty. Looking around at a house I thought would always feel big and empty, and then down at my ring finger, I felt grateful. Picking up the invitation again, I read and re-read the envelope, personally addressed in swirling calligraphy to Dr. June Scobee. I rubbed my fingers across the raised emblem of the United States presidential seal. I marveled at the long road I'd traveled to get to this point. Who would have believed that I, a child from the poorest family among poor neighbors, could have grown up to experience such a marvelous opportunity?

Did the president and his staff really know me, the sad secrets of my youth? The FBI had investigated my past and questioned several of my friends to learn about me as a prerequisite to giving me a presidential appointment that also had to be approved by Congress, but that investigation and those questions looked only into my life since I had been married as a teenager.

Anxiously, I returned my attention to the invitation, replaced it in the envelope and checked the clock. Because I'd dressed an hour early, I still had thirty minutes before the driver came to my home to drive me to the White House to be seated with presidents, royalty, and dignitaries. I laughed at myself—at the unlikely girl to experience such a storybook fantasy.

A CINDERELLA EVENING

I glanced down and saw that, as if a fairy godmother had visited, I was dressed in a lilac-colored gown with soft shades of pastel beads on the bodice. Fidgeting with my unruly curls and then giving up on arranging them, I reached into my tiny clutch bag, too small for anything other than a lipstick and compact, so I could take a peek into the mirror and dust powder on my nose. I fretted over my reflection, wondering if my makeup hid any signs of apprehension. Studying my image, I asked myself, "Is this really the same face of the misfit child who for years was evicted or homeless or lived scattered about with relatives or in a foster home?" I remembered the kids who taunted us and refused to sit next to us on the school bus: *The Kent kids stink! They ain't got no kitchen sink!* Was that bedraggled teenager the same person who tonight would sit at a table with kings and a president? As I thought of the upcoming evening, I grew even more nervous, fidgeting and pacing. Then the doorbell rang. Opening my front door, I saw my handsome prince, my general, standing there in his formal uniform. After a quick kiss, we headed to our waiting coach.

Within minutes, we were escorted through the lovely halls and rooms of the White House until we arrived at the reception and receiving line. Don and I stopped to study the seating chart.

"Tell me who will be seated next to me and something about them," I urged.

Don laughed when he found my name. "If you know anything about Egypt, you'll be fine. You are sitting next to President Mabarak himself and Barbara Bush. Also, Lee Iacocca will be at your table. Next to him is the singer Maureen McGovern."

"How about you?" I asked. We studied the chart and saw ten tables with ten people seated at each one. Don found his name and saw that he would be seated with Secretary of Defense Cheney and several senators and their wives.

When we arrived next to the president, his aide announced, "Dr. Scobee and . . . her general." Confused, he tried to correct himself. "Er, General Rodgers and Mrs. . . . ." The president heard the confusion, took my hand, and welcomed me to the White House, saying, "Hello, June Scobee. And this is your general? Hello, General. Nice to meet you." When we met Barbara, she delayed us for a minute, saying to Don, "We love this lady; you take good care of her!"

"Yes, ma'am!" Don responded gleefully.

Seated at the table, I looked across and recognized Mr. Iacocca and Ms. McGovern. Next to me sat the honored guest, President Mabarak of Egypt, his associate, and First Lady Barbara Bush. I tried to manage my anxiety by studying the place setting of elegant dishes and glassware, the white menu with gold lettering, and the name cards.

The president spoke a few formal words of welcome and acknowledged the special guest from Egypt and his wife, who was seated with him at the president's table. Smiling graciously, he said, "Enjoy your dinner."

I waited for Barbara to begin eating before I picked up my fork, but before she did so, she looked to her other table guests and said, "Let me introduce you to June Scobee. She was the widow of the *Challenger* commander." Without hesitation, she entreated, "June, tell everyone how you've helped the other shuttle crew members' families to build the Challenger Center for children."

I was accustomed to giving fifteen-minute talks, but I also knew how to tell my story quickly, having practiced speaking to CEOs of corporations. I gave the other guests an abbreviated overview, which only encouraged them to ask more questions. Eventually, I was able to slip in a quick question about Egypt, which carried us through interesting table discussions for most of the evening.

Along with the dessert, champagne was served. When Barbara mentioned that I was engaged to General Rodgers, I pointed him out across the room. It was easy to spot him, as he was the only guest wearing a military uniform. About the same time we looked to them, his table looked up toward us. Someone at his table lifted a glass, toasting in our direction, and then both tables gestured to the other in raised glasses.

After dinner, the entire group gathered for entertainment in the Blue Room. Maureen McGovern sang several songs, the most memorable being "There's Got to Be a Morning After." Following the entertainment, guests began leaving. Even after our delicious meal, Don said he was still hungry on the way home, and he stopped for a burger at a drive-through restaurant. I teased that he was such a little boy, and I hoped he would always remain so because I never ceased laughing at him or with him.

THE FAIRY-TALE WEDDING

On the evening prior to our June 3 wedding, my dear friends, Rose and Bill Narva at the Jefferson Hotel, hosted a wedding rehearsal dinner for the wedding party and our out-of-town guests. It was a joyful reunion. Having our children and their spouses with us, expressing their happiness for our marriage, meant the world to me. It was also wonderful to see Barbara Morgan and her husband, Clay. It seemed such a long time ago that she had served alongside Christa McAuliffe in preparations for her space journey. The evening was filled with laughter, loving stories, and a beautifully prepared delicious meal.

The following day, our family and friends attended our wedding in the small military chapel next to Arlington National Cemetery where both Dick and Faye had memorial services. I stood at the front entrance with my son, Rich, and looked longingly toward the man whose arms I would walk into for the remainder of our lives. He was truly a knight in shining armor, dressed in full uniform. Standing beside him as best man was his son, Eric. To the other side were Kathie, my matron of honor, and her four-year-old, Justin, holding a satin pillow with golden wedding rings.

Don's family and friends from his military career sat on the right side of the church. In the front pew on the left sat all seven of the families representing the *Challenger* crew: Dick Scobee's parents, Marcia Jarvis, Chuck Resnik, Cheryl McNair, Jane Smith, Lorna Onizuka, and Steve McAuliffe and his two children, Scott and Caroline. We also welcomed friends from Houston, the Challenger Center, and the White House.

The soft organ music ceased, and then the wedding march began, sounding like trumpets heralding a magnificent, triumphant event. My son took my arm, smiled, and walked with me down the aisle to a new life—two lives, two families, two sets of friends coming together to walk a new journey.

Chaplain Art Jensen led the ceremony. Jane Mahaffey, who had introduced Don and me to each other, read 1 Corinthians 13 from the family Bible. As we said our vows, our two sons each took a candle representing the Scobee and Rodgers families, and together they lit a single candle that represented our unity. The chaplain led us through our vows, then added, "We all come together united in spirit this day just as we will join together again someday with Dick Scobee and Faye Rodgers, who we know are together smiling down on us."

That day, I discovered joy as great as my sorrow had been deep. God blessed me with another chance to love and be loved, but more important was my rekindled spirit. I learned that the true key to happiness is when we unlock our hearts to give and receive compassion.

We enjoyed visiting with our guests at the reception. They had traveled from Washington, Texas, Florida, New York, California, and other places in between. Even my childhood friend and my daughter's namesake, dear Kathie from Florida, was there. All the Scobees congratulated us, and we had pictures made with the *Challenger* families, with whom Don also quickly bonded.

That night we slipped away to the Inn at Little Washington for dinner, and then on to a nearby bed and breakfast for one night before flying out the following day to our honeymoon in Bavaria, Germany.

HONEYMOON HOLIDAY

Bavaria was beautiful and romantic. Its history was intriguing, and the people were delightful. We saw castles, cathedrals, mountains, and rivers, and enjoyed opportunities to listen to nostalgic German music.

At one of the castles, I stood mesmerized, looking up at its magnificence. I figured King Ludwig's Neuschwanstein Castle must have inspired Walt Disney's Magic Kingdom. I felt like a true Cinderella. My prince showed me his personal favorites at Heidelberg, the ruins of the castle there, and Philosopher's Way, a walk university professors often take. We traveled the path together just as we would in life. I whispered a prayer, thanking God for the joy and many blessings.

We walked along the path, talking about our future but also about our loved ones in heaven. The Bible promises there is a heaven, and I told Don, "I tell my grandchildren that the soul is the part that goes to heaven when the body is left behind buried on Earth."

"Yes," Don agreed, "and we say things like 'heart and soul' or 'soul mate.'"

We walked along, "soul searching" together, and I recalled these words from Emily Dickinson's poem: "I never spoke with God, / Nor visited in heaven;/ Yet certain am I of the spot / As if the chart were given." Don and I walked on in quiet reverie and gentle contentment.

HOME TO FT. HUACHUCA, ARIZONA

Following our honeymoon, we returned to Fort Huachuca to the joy of a domestic home life. After the *Challenger* accident, I had thought I could never be happy again. Now, I was in love with life once more.

The Officers' Wives Club invited me to a reception of welcome. During the gathering, they asked me to step outside the building to a lovely spread of lawn that reached out to the desert. As we waited, I saw a swirl of dust moving toward us. As it came closer, I distinguished men on horseback. Soon, I recognized that they were soldiers representing the early Army Cavalry. When one of the troops dismounted, he handed me a beautiful bouquet of yellow roses.

"Our troop welcomes our new First Lady to Fort Huachuca," he said gallantly. Then he climbed back on his horse, and the group turned to ride into the hills from which they came. I was delighted with the surprise.

With such a thoughtful welcome, I found it easy to settle into my new life. On weekends, Don and I traveled the area as tourists. We visited the historic Old West towns of Tombstone and Bisbee. A few times, we drove into Tucson for a movie or special dinner. Sometimes we rode a motorcycle around the scenic countryside or along the famous winding Route 666.

Early in the summer, Uncle Gordon, who had cared for my brothers and me during difficult years of our youth, came for a visit. Now an aging veteran with white hair like our grandfather, he relished the nostalgia of the historic army fort. Later, all three of our children and their spouses visited at the same time. Kathie had two children, and Rich had one. We enjoyed having them in our home, the grand old-world place with five bedrooms. A photographer took our picture posed sitting together on a cannon out in front of the house.

Watching Don Rodgers as he interacted with our family and friends, I learned how to cultivate a sense of humor. He balanced me well and taught me how to take myself less seriously. If he wasn't whistling a tune while he worked, he was thinking up a prank or joke. I appreciated Don, our strengthening bond, and the path we were creating to our future.

AN UNEXPECTED GUEST

Our family had to leave before the Fourth of July celebration. I was disappointed that they couldn't stay for the parade on post, but I looked forward to participating in the military pageantry with Don. Don dressed in his uniform to ride in the parade, and I wore a white linen dress. Just as we were driving from our home to the parade grounds, Don received a call from the army post hospital informing him that former President Reagan had been admitted.

We drove quickly to the hospital, met with the doctors, and learned that Reagan had fallen or been bucked from his horse while riding on a nearby ranch. Don met with his hospital staff while I waited. Soon, Don returned to me, saying Nancy Reagan wanted to see me. When I saw her, she looked anxious and worried about her husband.

Because it was a holiday, the army physicians had difficulty locating Reagan's baseline medical records. Having lived for such a long time at the Jefferson Hotel in DC, I knew that Rose Narva's husband, Admiral Bill Narva, had been the attending physician for Congress.

After retrieving the former president's records from Bill, the physician invited Nancy to join her husband. She turned quickly to me and asked to borrow my lipstick. Dressed in jeans and a western shirt, she explained that she was working on a manuscript and had not bothered with makeup. Of course, the former Hollywood actress still looked lovely.

With fresh lipstick, she stepped into the hospital room for a few minutes to see her husband. Soon she returned and motioned for me to join her. "Ronnie wants to see you," she whispered.

As I entered the room, the former president thanked me for helping take care of his Nancy. Don followed me into the room in time to hear Reagan say, "I've been watching the television, and the news reporters are saying I fell off my horse. It's not true. Let them know I was *bucked* off my horse. We had a one-man rodeo going."

The physician came in and said they had more tests to run on Reagan. He would hear none of staying in the hospital, though, and continued to gather his things. Sensing everyone's anxiety everywhere, I asked if the hospital could send a physician on the helicopter to assist Reagan. They agreed, and everyone was relieved. We watched them climb aboard the helicopter and fly south toward the ranch. Don looked at his watch and realized the parade was long over, so we drove home and relaxed with a backyard picnic.

THE BERLIN WALL FALLS

A marvelous bit of world history took place while we were stationed at the Army Post in 1989. On the November 9, the Berlin Wall between East and West Germany fell, bringing the end to German separation. Watching the TV broadcasts, it was great to see the joy expressed by presidents Reagan and Bush. The German officers stationed at the army post were so elated that they provided a reception so we could all help them celebrate. When some of Don's soldiers returned from Germany after the historic event, they presented Don with a chunk of the wall mounted on a plaque.

MORE SPACE ADVENTURES

During our first year of marriage, I stayed involved with the Challenger Center education programs long distance, but continued to travel to Washington, DC, for quarterly board meetings. At the spring board meeting in April 1990, we had reason to celebrate. The STS-31 *Discovery* shuttle flight launched the Hubble space telescope. Led by Commander Loren Shriver, the crew, all dear and personal friends, launched the Hubble into orbit 347 miles above Earth. NASA planned for it to capture spectacular images of distant galaxies, helping us uncover the mysteries of our universe.

Always, it was wonderful to see friends during times like these, especially the other spouses of the *Challenger* crew who continued to serve on the board of directors with me. Our paths ran parallel for the first few years, but ultimately they veered off in individual directions, intersecting periodically for loving reunions.

Chapter 19

STAR-BRIGHT EXPECTATIONS

DON'S NEW ASSIGNMENT

Don's tour of duty as commander of Fort Huachuca ended the following summer in 1990, when he left the Arizona post to take an assignment back in Washington, DC, as director of the Defense Communications Agency. With the developing Gulf crisis, some of our nation's military was shipping out to the Middle East. It was an incredibly tense time for my husband and the country. Earlier that year, the Iraqis had held hundreds of Americans hostage, and the stock market was volatile.

We moved into Quarters 4 at Fort McNair, an antebellum four-story military home across the Potomac River from the White House. Fort McNair was the headquarters of the U.S. Army Military District of Washington. The house offered front and back porches and six bedrooms, with enough space for several families.

After we settled into the house, I talked with Eddie, Don's aide, about its beauty, size, and history. Growing up in poverty, I had never imagined a house like this one, nor in my wildest dreams had I thought I would live in such grandeur. Eddie reminded me that with Don's new position, we would need to entertain visitors and dignitaries for business and social events. Thus, the formal living and dining areas downstairs were kept spotless and presentable for special visitors, and the upstairs bedrooms and sitting areas were reserved for our private and personal attention.

One fall afternoon, I returned home from working at the Challenger Center headquarters in Old Town, Alexandria, Virginia. Eddie met me at the door, poured me a tall glass of iced tea, and then told me about potential issues with the other military wives. It was time for the Signal Officers' Wives Tea, and as the Senior Signal Officer's wife, I was expected to have the tea in our home.

"The problem," Eddie explained, "is that some of the wives don't think you should hold the place of honor since you've been married to the general for only a year, and they still love the other Mrs. Rodgers, the general's first wife."

OVERCOMING REJECTION

It is often more difficult to be a remarried widow than a remarried widower. The man returns to his job and keeps his name and identity. The woman takes her new husband's name. "What happened to June Scobee?" I wondered. "Who is she now? Who is Dr. Scobee, the career woman?" I resolved that I was neither the same bewildered person of my youth, nor only the widow of Dick Scobee, nor only the new Mrs. Rodgers. I was *all* of these people. I had overcome misfortune and enjoyed great experiences, and if I worked up the courage, I could find confidence to meet the wives who loved the first Mrs. Rodgers. I believed in myself, and I felt certain that God would help. I just needed to work on my attitude.

That night as I spoke with Don, I told him about the struggle with my identity and name. I wondered if the name "June Scobee Rodgers" would be appropriate. Don encouraged the full use of my name—Dr. June Scobee Rodgers. Later, I asked to drop the Dr. prefix that seemed too pretentious or at least too big a mouthful. Don teased me good-naturedly about my identity struggle.

Eddie and I began preparing for the tea, and I requested Faye's and my tea sets, as well as my grandchildren's miniature doll tea sets. We placed the large sets in groups around the banquet-sized dining room table, and arranged the toy sets on small stands as centerpieces. We also included flower arrangements and, in front of each tea set, a place card with the name of the tea.

On the afternoon of the tea, nearly all the wives came to our home. I welcomed several dozen ladies who entered the house cordially and quietly, then nibbled on petite finger sandwiches and Eddie's famous lemon tarts and shortbread cookies. Eddie kept the kettle boiling with water to refresh the tea as the ladies greeted one another and socialized.

I asked to speak, and the room grew uncomfortably quiet. Smiling, I reminded the ladies about the joy of childhood tea parties. They nodded and smiled at their own memories. Then I appealed to them to share their stories with me. "All of you knew and loved Faye," I said, hesitating as tears brimmed from my eyes. "I want to love Faye too. Will each of you take a turn to tell me one of your favorite memories of Faye? And if you see an item in the house that reminds you of her, please share that story too."

Their stories filled the room with laughter, and the sensitive memories of their loss added a somber quality of respect for the former Mrs. Rodgers. Surrounded by her friends, I felt close to them and also newly fond of the woman who was Don's wife for more than thirty years. At the end of the tea, I waited at the front door to shake the women's hands, but one by one, they hugged and thanked me instead. I was relieved and grateful for the success of the afternoon.

THE MIDDLE EASTERN CONFLICT

That night, my mood shifted from pleasure to tension as Don told me he would be preoccupied with preparations for Desert Storm in the Middle East. He needed to fly to the desert to meet with four-star general Norman Schwarzkopf, commander of the Allied Forces, whom I had met just after marrying Don. A big, burly man, Norm had given me a bear hug while Don took a photo of us.

Don explained that we would need to stay in DC for the upcoming Christmas holidays. The military was on alert, and great masses of troops were being sent overseas. Within days, wives of the officers from the Pentagon set up our own telephone command post to answer calls from the younger wives, keeping them posted on the whereabouts and safety of their husbands or other family members in the desert. I also served as an Arlington Lady to represent the commander when he could not attend a funeral. On these occasions, I presented the folded flag to the families who buried their loved ones at the cemetery.

Christmas rushed in without our traditional celebration. Don and I slipped away to the historic Red Fox Inn near Washington, DC, for an overnight stay, a quiet Christmas Eve dinner, and a Christmas Day exchange of small gifts. We walked around Middleburg, and then returned home for the serious business at hand. The news was sobering about the Persian Gulf and Saddam Hussein, who had taken over Kuwait. My husband was so busy that I never knew what time of day or night he would be able to come home, so I kept a pot of steamy soup ready on the stove for him.

For me, Operation Desert Storm began on CNN news reports. Pilots were captured. I worried about my son, Captain Rich Scobee, who was stationed at Ramstein Air Force Base in Germany. I didn't know if his squadron was flying their F-16s over Iraq at the time, but later that year, Rich gave us a picture taken of him flying over the area that Don's former command, the 7th Signal Brigade, had named Camp T. D. Rodgers.

In late February, President Bush announced in a talk to the nation that after forty-two days, he was calling for a ceasefire if Hussein accepted the United Nations resolutions. As we waited anxiously for good news, one bright spot was when Don and I celebrated our first meeting at the sunrise service at Arlington. I invited two-year-old Emily June, our sweet granddaughter, to stay with us in the big house, attend Easter services, and find colored eggs on our sprawling lawn. We were even invited to the White House Easter Egg Roll. Emily was a precious distraction during the otherwise tense time.

Later in the spring and early summer, our troops began coming home, and even the news media reported that we had won the war and could take pride in how quickly it was fought. The famous Bush couple attended heartwarming events at Arlington National Cemetery for the families and soldiers who had died. It reminded me of the day less than five years earlier when they attended a ceremony with the *Challenger* families as we buried our own loved ones' remains at the memorial near the amphitheater.

OUR RETIREMENT CITY

When the Desert Storm victory was declared, Don talked with me and then announced to his staff that he planned to retire from the army in September 1991. He wanted us to ride out into the sunset on his motorcycle after serving his country for more than thirty-four years. It was time to move to the next phase of our lives.

Finally, no military assignment would dictate our next place of residence. Don, being an engineer, decided that since I didn't agree to move to the Arizona desert with him, and since he didn't want to live in DC with me, we would create a matrix, placing the names of ten cities down one side of the paper and listing five qualities of life across the top. We took turns adding potential hometown names to the list of our favorite cities, including

Charleston, Savannah, San Antonio, Augusta, DC, Tucson, and Chattanooga. Across the top we listed qualities like having four distinct seasons, a college town, outdoor sports, a symphony, and a regional airport.

I wrote to the chamber of commerce for each city and collected information. Separately, we filled out the matrix checklist with each of our choices, and then we added the score and compared our totals. Surprisingly, Chattanooga, Tennessee, came out on top. Both of us had grown up near the area, but neither of us had lived there, so we decided to visit the city and have a look for ourselves. In the valley between Lookout and Signal mountains, we walked along the river, met friendly people everywhere, and fell in love with the interesting history and special beauty of Chattanooga.

Soon we found acreage to purchase on the side of Signal Mountain and began looking at house plans so we could build our retirement home. We agreed that it had to be big enough for all three of our children and their families to visit at the same time. We chose to build an English manor. The contractor we hired began construction as Don and I returned to DC for his retirement ceremony in September.

CHALLENGER CENTER CELEBRATES FIVE YEARS

The Challenger Center experienced tremendous growth and change in its first five years. We had several dedicated and hardworking presidents; before Jim Rosebush moved on, he introduced us to David Winstead, who brought a significant seriousness to the organization, hired the right people, and worked with the Houston Museum of Science to create our first Challenger Learning Center (CLC). David was succeeded by Doug King, who rolled up his sleeves and began the vigorous growth of the organization, creating one center after another. He was dedicated and serious about the responsibility of his position, initiating fundraising opportunities and special events.

One event involved asking Barbara Bush to visit our local Owens Science Center CLC in Greenbelt, Maryland. We planned to celebrate our fifth year as an organization and have Barbara cut a birthday cake with five candles. First, Barbara visited the simulator to watch the students on their mission to the moon. Reporters and cameramen stood at the edges of the room, their backs to the walls and cameras rolling. Not one child looked up to greet Barbara, so one of the cameramen asked, "Hey, kids! The First Lady is here. Won't you say hello?" No child noticed. Finally, with prodding, a girl at the navigation console placed her finger on the computer screen, looked up at Barbara, and said, "Oh, hello, Mrs. Bush. I can't stand up to meet you because I might lose my place, and we'll be lost in space."

Later Barbara Bush told me, "June, what a wonderful place! It was full of knowledge, enthusiasm, and fun. Through the Challenger Learning Centers, now opening all over the country, you have given the children of our country a great gift: the gift of hope and the gift of learning."

When the First Lady stepped into our CLC conference room to cut the cake, Secret Service members surrounded us and told us we could not use a sharp knife to cut the cake. We scurried around and found a dull plastic knife to do the job. Barbara blew out the candles, said a few words of congratulations, and sliced the cake. Gobs of sticky frosting clung to everything she touched, but goodheartedly, she licked the sweetness from her fingers, and we continued with the celebration.

Barbara Bush is always gracious, charming, and delightful. She's been a dear friend, joining us with her husband when he was vice president and again as president to somber memorial services and special Challenger Center events. I adore her not only for her kindness to me, but also for her lovely sense of humor. She is *real,* a significant compliment in my estimation for a woman of her prominence.

DON'S MILITARY RETIREMENT

Within days, arrangements were made for Don to retire. His son Eric, Eric's wife Anne, and their baby, Margaret, came to the retirement service. Eric was proud of his father and wished his mother could be there to help him celebrate, but nothing dampened the celebration of Don's service to his country for thirty-four years.

I had grown into my own identity as well and learned to love us being introduced as Lt. General and Mrs. Rodgers. No longer was I the "new" Mrs. Rodgers, and usually I insisted on being called June. I was proud to be Don's wife and to move into retirement with him. We looked forward to all kinds of new adventures.

On one such adventure, we traveled to the four corners of the country on a Goldwing motorcycle, stopping in various locations to visit friends and family. We each packed a tiny bag of clothing and stayed at cheap motels—quite an experience for me after the life to which I'd become accustomed.

Along the way, we had wonderful visits with Steve McAuliffe and his new wife, Kathy; Clay and Barbara Morgan; with Dick's parents, who had claimed me as their daughter and Don as their new son-in-law; and with my dear friend Kathy Casper, with whom I had taught so many years ago at Edwards AFB. We spent time with Marcia Jarvis, whose husband, Greg, had been the mission specialist on the *Challenger* flight. She was also a fellow board member of Challenger Center, so we had much to talk about. In California, we saw our dear friends, the Narvas. Bill had retired from the Navy, and they managed a lovely hotel in Palm Springs. We continued on to see my daughter, who was working in public relations for the College of Law in Houston.

Our trip had stretched from Tennessee, along the Blue Ridge Parkway, on to DC, New Hampshire, Idaho, Washington, California, Arizona, and Texas. As the weeks went by, I grew weary and joked that I was also getting more bow-legged every day that the trip stretched on. Along the way, people commented that I must really love Don to take such a journey with him. They were right. Still, I would not have traded the experience for anything. I had been tired and dirty, but I had also been exhilarated and inspired.

CHALLENGER CENTER GRAND OPENING

Don and I settled in at Chattanooga. Over the next months and years, the children and grandchildren all visited during holidays and for summer vacations. We enjoyed visitors to our home.

I also traveled frequently to our Challenger Center office in the DC area and to other areas to help communities celebrate the grand openings of their learning centers. Larry Adams from Martin Marietta Aerospace replaced Manny Fthenakis as our chairman (the chairman leads the board and helps direct the president), serving as a magnificent fundraiser. Vance Ablott also left his position as Director of Space Center Houston to become president of the Challenger Center. He was a delightful, creative engineer who continued to grow the Challenger Learning Centers at leaps and bounds and organized a large team of employees, purchasing a building to contain all the activity.

Meanwhile, NASA successfully flew one shuttle after another, launching satellites, repairing others, and conducting research investigations in the weightless environment. Eileen Collins, NASA's first female shuttle pilot, flew aboard *Discovery* on STS-63. In June 1995, a few of my dear friends flew on *Atlantis* STS-71 to rendezvous and dock for the first time with the Russian *Mir* space station. The crew commanded by Hoot Gibson returned with my friend Norm Thagard, who had launched into space with the Russians. Soon after his homecoming, Don and I visited with him and his wife Kirby, and I asked what he did for entertainment while living on *Mir*.

He teased, "We watched American videos dubbed in Russian on a Japanese VCR." When Norm retired from NASA as a mission specialist and medical doctor, he accepted a position at Florida State University in the College of Engineering and built a Challenger Learning Center with an IMAX Theater across from the capital in Tallahassee.

In Chattanooga, I wanted to keep a low profile, but eventually, our congresswoman, Marilyn Lloyd, asked me about building a Challenger Learning Center at the university in Chattanooga. Don encouraged the idea because he knew how the simulation activities could motivate students to study math and science. I sought a co-chair in Marcie Pregulman, and we agreed to find a team to join our effort.

Fundraising to build a local center versus a national organization was different, but university president Fred Obear, professor Bernie Benson, and I joined to organize local leaders in government and business to create a center on the university campus. Hardy Caldwell's Family Foundation provided funds for the College of Engineers' senior class project to design and build a shuttle for the center. John Germ, a local engineer, gathered architects to create the design, and John Guerry of Chattem Industries led a skillful fundraising campaign. Friends at the McKee Foods Company agreed to provide funding for the simulators.

The twenty-fifth completed center was built dedicated to Ruth McKee, who had never fulfilled her dream of teaching school due to her focus on the food business. I was happy to announce to all the visitors at the grand celebration that Ruth McKee's classroom space simulator doors were open to children. To help me celebrate, our three-year-old granddaughter Margaret Faye Rodgers helped me cut the cake. My daughter, Kathie, joined us from Houston, and we all celebrated in fine fashion with Don's son, Eric, and Eric's wife, Anne.

INTERVIEW ON THE *HOUR OF POWER*

A turning point for me came when I met Helen McDonald Exum, whose family owned the *Chattanooga Free Press*. She introduced me to Robert Schuller, who invited me to the Crystal Cathedral in Los Angeles for an interview on the *Hour of Power* program that was telecast around the world.

There in front of the TV cameras, Dr. Schuller asked me to tell my story about the *Challenger* accident and about creating the Challenger Learning Centers that he referred to as a "triumph over tribulation." He also challenged me to write the story in a book and then return to his program to talk about it. When I got back home, I asked Don what he thought. Excited, he urged me to write a book in time for the tenth anniversary of the *Challenger* accident, which would occur two years later in January 1996. It was an opportunity I decided to seize.

Chapter 20

SILVER LININGS: TRIUMPH OF THE *CHALLENGER* 7

TEN YEARS

In 1986, our nation was united in disbelief and great sorrow, but because we triumphed over our tragedy, ten years later we united again in celebration for a decade of reaching, teaching, and touching the future. On April 11, 1995, as we approached our tenth year of continuing the education mission, the *Challenger* families gathered in Washington, DC, with our friends, including George and Barbara Bush, to recognize our accomplishments.

Since the day he first joined us at the crew quarters at Kennedy Space Center in Florida, President Bush has kept his promise to answer our call when we needed him. During the ceremony, we presented the Challenger Center Presidential Award to the Bushes.[1]

TELLING THE STORY

Beginning the process of creating my book, I talked to others who had traveled the journey with me and took notes. I collected photographs, press releases, meeting minutes, and newspaper articles. When I felt ready to start, I wrote for hours about the events of the tragic day and the sad and troublesome times that followed. Reliving the days brought back many memories, but hidden in the sorrow I discovered delightful recollections of Dick, our children, and our friends. I cried and heaved great sobs for the traumatic vision of the shuttle explosion and the memory of the shock that followed, but I also treasured the joyful memories that surfaced.

At times, I was so engrossed in the writing that I felt as though I were reliving the day, but looking up to the sky, I discovered that I had found not just one but many silver linings beyond the dark clouds. The writing of the book was a journey in itself—like a trail into the wilderness. At the beginning of the trail, my daughter helped me plan, gave direction, and offered encouragement. Don was there as needed to keep me motivated.

Near the end of the trail, Kathie and Rich read the manuscript and encouraged its publication. Their tender words of support made all the difference. At the end of the journey, I felt grateful for their help. I was also thankful for Don's patience and assistance in seeing the work completed, and for his understanding and compassion on our own journey together.

By fall, the publisher Smyth & Helwys of Macon, Georgia, published the book that we titled *Silver Linings: Triumph of the Challenger 7*. The story was featured on news and morning television programs and in magazines and newspapers. As promised, Robert Schuller invited me back to Los Angeles to talk about the book his *Hour of Power* Sunday program. I was pleased to return and share about my personal revolution. And at every interview, I proudly announced that my son would fly a tribute over the football stadium for his father on Super Bowl Sunday.

SUPER BOWL HONOR

That year, January 28 fell on Super Bowl Sunday. The Dallas Cowboys were playing the Pittsburgh Steelers at Sun Devil Stadium in Phoenix. Ten years earlier, in 1986, seven astronauts had died on that day. As part of the pre-game show, my son was invited to fly his F-16 jet in the missing man formation over the stadium. Captain Rich Scobee and his fellow pilots from Shaw Air Force Base flew to Phoenix in honor of Rich's father.

The largest television audience in the history of football tuned in to watch the game. Standing at the 50-yard line, Don and Kathie and I held each other as the formation flew overhead and Rich turned his plane upward to the heavens. The announcer explained that the missing man formation was flown on the tenth anniversary of the loss of *Challenger* and that Captain Scobee flew in honor of his father and the crew. We looked to the heavens to see a son honor his father and imagined a father looking down at his son.

Everyone cheered! The honor was a glorious tribute to the *Challenger* astronauts, but for me the thrill was in seeing my son have the personal pleasure of honoring his father with such a momentous opportunity.

The Cowboys won, 27-17, and the following morning, an editorial appeared in the *Arizona Republic*:

> It was the single-greatest moment in Super Bowl history. It wasn't just about sports and entertainment and money, not for this moment. It wasn't team against team, player against player, or city against city.
>
> The single-greatest moment in Super Bowl history—when the planes flew over the stadium. After the Star-spangled Banner, the crowd was still quiet, then four F-16 jets appeared on the southern horizon.
>
> The flyover had been planned to commemorate the 10th anniversary of the space-shuttle *Challenger* tragedy, which killed the first teacher in space, Christa McAuliffe, and six other astronauts. It had been announced that the formation would be led by Capt. Rich Scobee, son of the *Challenger* Commander, Dick Scobee.
>
> It was the game that got our attention. It was the game that caused so many to travel to Arizona and tune in on television, but it was the moment that had value, it was worth more because it had more to do with honor and heroism and memory.
>
> Then the planes appeared. We watched them fly in, watched the people wearing black and gold and the people wearing blue and white look up as one at the sky. It didn't last long. They performed what is called the "missing man" formation. The planes approached in a tight group until they were just above the stadium. Then one jet broke away from the rest. That was it. The son of an astronaut flying straight toward the heavens in honor of his father and crew.
>
> A father looking down on his son.[2]

As Don, Kathie, and I entered the airport to return home, a large group of people came up to meet me. I thought I was finally known not as the widow of the *Challenger* commander but as the author who promoted *Silver Linings* on CNN, *Good Morning America*, or *60 Minutes*. As they approached, I smiled a greeting, only to hear them ask excitedly, "Are you the *mother* of the F-16 pilot who flew over the Super Bowl?" Laughing at myself, I nodded proudly.

Inside the B 737, I looked to the clear blue sky and asked God to help me to keep my life in perspective. With such a public story, I wanted to maintain a sense of humor and humility. I hoped that knowing about the loved ones left behind and what we accomplished could help others realize that the endurance of hardship builds character. It certainly built mine. Through the darkest of clouds, God led me to see the silver lining.

Words from my childhood surfaced as the plane took off. When I was young, my mother had read A. A. Milne's *Winnie the Pooh* to me. In one story, Christopher Robin said to Pooh, "Promise me you'll always remember: You're braver than you believe, stronger than you seem, and smarter than you think." I've never forgotten this advice.

Notes

1. The Bushes' remarks on receiving the award can be found in the appendix.

2. E. J. Montini, "A Moment of Pre-game Greatness," *Arizona Republic,* 29 January 1996.

Chapter 21

SUNRISE, SUNSET

TAKING THE NEXT STEPS

With my story told in print and the interviews behind me, I wondered about my next step. It came quickly, as a publisher invited me to create a coffee table book about Chattanooga, my hometown. I immersed myself into this project, which I felt was a fitting way to show my love for the city. After completing the book, I participated in local efforts, serving on the Board of Directors for the Thompson Children's Hospital and the National Center for Youth Issues. I responded to speaking engagements for local civic organizations, churches, schools, and youth groups. My days grew comfortable. Don and I entertained in our home and arranged for special events for the symphony, our church, and the university. We even co-chaired citywide fundraising events.

I thoroughly enjoyed watching my children's careers expand. Rich worked his way up as a squadron commander, then wing and base commander, and Kathie moved to Chattanooga to work in public relations for the Tennessee Aquarium. Their families grew, and Don and I treasured every minute with our eight (and counting!) grandchildren. One Mother's Day was particularly special, as Kathie gave me a framed poem she had written for me (included in its entirety in the appendix), which read in part,

> . . . God made Muse Mothers to love and inspire
> To whisper muse words and
> Muse music inside us.

To teach us the magic
In the dance of our lives
In the faith of the seasons
In the rhyme of God's sighs . . .

Her words still encourage me when I feel inadequate or insignificant. I've learned that mothering our babies and children even as they grow into their adult lives must certainly be the most creative, most challenging, and also most rewarding gift God provides.

SAYING GOOD-BYE TO MOTHER

Once she received a proper diagnosis and appropriate medication, my own mother had overcome many issues related to her mental illness. When she was physically able, we enjoyed her visits in our home. Regularly, my brothers and I visited her at her San Antonio apartment. She was generally delighted to see us and always hospitable, but we worried as her smoking increased and she grew more frail. My brothers took turns driving her to the hospital, and I made a special effort to travel at least once a month from Tennessee to see her.

Despite our pleas, she refused to stop smoking, and I gave up the quest when she reached her seventy-ninth birthday. One day, a nurse in a San Antonio hospital called and said Mother was suffering from cardiovascular disease and wanted me there with her.

My brothers and I were at her side as quickly as possible. The doctor told us Mother would die soon, and advised us to make her comfortable, possibly with hospice care. Mother could barely talk, but she refused to move in with any of us. She insisted on going to a particular nursing home, so once her strength returned we helped her settle in.

The nursing home attendants became impatient with her. Attached to a tank of oxygen, she'd light a cigarette. She transferred to several different homes over a period of two months. If she couldn't manage a cigarette, she'd complain of terrible pain, so an ambulance would take her to the hospital, where she'd find a stranger to give her a cigarette. The attendants at one nursing home discovered her ruse, so she started a fire in her bathroom sink to set off the fire alarm. When all the residents were moved outside to safety, she found another cigarette.

Finally, in February 2001, Mother was admitted to the hospital for what we were certain was the last move in her transient life. In her hospital room,

I pulled a chair close to her bed. She pointed to a small box, and inside I found her prized pearls and a delicate hand mirror she had always treasured.

Breathing heavily, she said, "Junie, it hurts so much to breathe and to talk." Gasping for air, she told me to take the pearls and mirror and also a cedar box. I held the box and studied the picture of an English pastoral scene laminated on the lid. It was a rare gift from Daddy, bought at a gas station on one of our moves from Texas to Alabama. I opened the box and discovered a collection of letters that I had written to her over the years.

Touched, I kissed her forehead and embraced her. I held her hand until she fell asleep and thought about her gifts. These were treasures of a lifetime, small enough to fit into a little box. The thought made my head spin with unresolved questions.

The following day when I returned from my hotel to her bedside, the hospice caretaker told me Mother was troubled about forgiveness. "Either she needs to seek forgiveness, or someone needs to forgive her," the worker explained.

During that visit, as Mother struggled for every breath, she and I talked about her years of anger. She was nearly always infuriated with a family member. I asked if she could forgive the people for whom she still harbored resentment. She shook her head no. I talked about how unyielding anger with someone ties us to the person rather than freeing us to move on. I even mentioned a story I heard about anger being like a splinter in your finger that festers and hurts until it is removed.

Mother cried. I prayed. I cried. I reminded her that Grandma and Papa were waiting in heaven when she was ready to see them, and that God had prepared a place for her too.

"You've been a wonderful mother," I whispered.

"I have?" she said, choking and wheezing.

"Yes," I assured her. "You loved us, taught us manners, cared for us, and read fairy tales to us when we were little. You were the best mother you could be, given all our troubles."

Mother cried and gasped for air. I asked the hospice nurse for pain medication, which seemed to soothe her even as she continued to suffer. When she slept again, I left her bedside to regain my composure. The hospice nurse later said Mother never awakened from that nap. She lived past the year 2000, which was something she had hoped to do. I prayed that she finally had her Easter morning.

My sisters-in-law helped make arrangements for Mother's funeral and burial. The family gathered at the burial site several days later. We brought pictures of her in her happier days—laughing with Mickey Mouse at Disneyland, balanced on Lee's motorcycle, standing with her grandchildren and great-grandchildren—and placed them on a pegboard. We each said a few words about her, prayed for God to take her, and sang a Bible song we remembered from our childhood. I felt blessed that my children had been able to know her, that they had heard her laughter and wonderful stories of her own parents and brothers. Kathie and Rich's compassion for her grew as they became aware of her mental illness, and I was grateful they could see her for the generous, interesting person she was.

There was no grand sermon and no gathering of multitudes of friends for Mother. A simple arrangement of flowers, a collection of photos, and her children and grandchildren surrounded her to bid her farewell. I truly believe Mother lived her life in the best way she could. She suffered for years with mental illness and its public stigma, and then she suffered physically with an agonizing disease that led to her death. But she left the world with four children who loved her, four children who discovered their purpose and contributed to the world in their own ways. She died with dignity, letting go of the bitter things and burdens and transformed into a truly loving being.

A TRAGEDY VIEWED FROM AFAR

Mother was born during the period between two world wars, during the "Golden Age" of aircraft advancements from the first bi-planes made from wood and fabric to sleek aluminum, high powered planes that barnstorming pilots like Amelia Earhart flew. Her life spanned the decades that saw advancements in transportation, communication, medicine and entertainment. She witnessed her son-in-law fly the shuttle, and saw space ships travel to the Russian space station *Mir* and the International Space Station (ISS).

The ISS is a joint space program begun in 1993 as a result of an agreement between President George H. W. Bush and Russian president Boris Yeltsin. Russia's aging *Mir* had served a tremendously useful purpose prior to breaking up in Earth's atmosphere and falling into the South Pacific in March 2001. On a clear night, the ISS is visible orbiting every 90 minutes in the sky, a tiny sparkle of fast-moving white light. Looking back from the ISS, astronauts tell us they can see all of Earth's features—both natural and man-made.

When our dear friend and astronaut Frank Culbertson commanded the ISS in autumn 2001, he recorded in his diary the magnificent views from the orbiting space station:

As we left the sun behind and the glare faded, the magnificent cities on the East Coast of the US magically appeared in the blackness below, outlining the Atlantic seaboard as clearly as if drawn on a map with a pen filled with glowing orange ink. From my perch over the North Atlantic—Boston, New York, Atlantic City, Norfolk, Charleston, Jacksonville—all could be seen lighting up for the evening as we sped out over the ocean and headed southeast. And hanging in the sunset above them were, I think, both Mercury and Venus.

Just after sunrise on the following day, September 11, 2001, Culbertson recorded another view seen from the ISS:

It's horrible to see smoke pouring from wounds in your own country from such a fantastic vantage point. The dichotomy of being on a spacecraft dedicated to improving life on the Earth and watching life being destroyed by such willful, terrible acts is jolting to the psyche, no matter who you are. And the knowledge that everything will be different than when we launched by the time we land is a little disconcerting.

On Earth, the news traveled quickly that tragedy struck our nation. My phone rang early. It was Kathie calling from her office, telling me planes had hit the towers of the World Trade Center and urging me to turn on the news. Along with the world, I saw the smoking towers, and learned of a third airliner flying into the Pentagon, just outside Washington, DC. Of course, we all later found out that a fourth had crashed into a field near Shanksville in rural Pennsylvania.

As the burning Twin Tower buildings crumbled, Kathie cried uncontrollably for those who were dying, those who were watching loved ones die, and for her own father, who had also died in a public tragedy. I told her to meet me at our church and we would pray together for the families. Our minister opened the doors to the church and had a special service of music and prayer for anyone who wanted to worship.

As the others sang softly along with the piano music, Kathie reached in her purse for pen and a scrap of paper and began writing furiously. She was writing a letter to the children. She wanted to let them know they would be okay, but she was not managing her own grief well at all. For many years I thought my children had overcome the loss and grief of their father dying so publicly, but I could see that was not true.

I drove Kathie home so she could send an e-mail to the children of the 9/11 disaster. Kathie grieved for her father and for all the parents lost in New

York City; in Washington, DC, at the Pentagon; and on all four hijacked planes. Like many people, Kathie needed many days to work through her grief before she was able to move forward in her personal life.

Famous newscaster Paul Harvey has said, "Storms are part of the year-in and year-out climate of life," but some storms are of such magnitude that they shatter the hearts of a nation and frighten our families. They may be ominous and dark, menacing and threatening like the despair of the 9-11 tragedy. Other clouds of a more personal nature may come in the form of great illness, death or divorce, fear or anger. But clouds are not forever. Just beyond the cloud is a bright light radiating with the warmth of God's love to help us weather the storm. Faith can turn the dark to light, and hope can triumph over tragedy.

FAITH AND FRIENDS

One way I triumphed in my daily life was by finding an opportunity to make friends. A group of us led by Fran Simmons organized a regular Tuesday lunch, gathering in some of our homes or at restaurants. Two of the women were nationally known speakers, one was a member of the family who owned the local newspaper company, one owned a department store, and one was a lawyer. Others had talents in interior design or philanthropy. I, the educator and founder of a nonprofit organization, was known as the "spacey lady."

One Tuesday as I hurried out the door to attend our luncheon, my husband asked, "Are you meeting with your Tuesday Floozies today?" I laughed and later shared the comment with my friends during lunch. The following week, one of our ladies, Nell Mohney, wrote about us in her weekly newspaper column, and from that time forward, people knew us as the Floozies. Our husbands, well-established leaders themselves, were dubbed the "Floozy Auxiliary." Together, we helped others, traveled, and had lively discussions about life, love, and tribulations. It was wonderful to have friends I could trust with confidences and who could celebrate each other's accomplishments.

VATICAN OBSERVATORY FOUNDATION

Don and I continued to travel together as well, visiting dear friends overseas and stopping by some of the Challenger Center international sites. On one trip, we visited Lindy Boggs, former congresswoman from Louisiana who had lost her husband in a plane crash and now served as ambassador to the

Vatican in Rome. She introduced us to the people associated with the Vatican Observatory located in Castel Gandolfo. We met Father Coyne, a Jesuit Priest and director of the Vatican Observatory. He led us on a tour of the library that contained a valuable collection of rare antique books, including works of Copernicus, Galileo, and Newton. I held the first-edition books in my hand and turned the pages, reaching back in time while also looking forward to the stars in the heavens. In addition, we saw a unique meteorite collection that scientists are using to learn about the early history of the solar system.

I learned that at the request of the Pope, astronomers at the Vatican Observatory had studied the universe and mapped the heavens down through the centuries. From its origins in the Gregorian reform of the calendar, the Vatican Observatory continues this legacy at the forefront of astronomical research. In addition, I learned that they have brought students and scholars from across the globe to the Vatican to study the universe in pursuit of their passion for astronomy.

Most striking, I found that the Vatican Observatory Foundation was established in 1986, the same year as the *Challenger* accident, and that it is located in Tucson, Arizona, where Dick graduated from the University of Arizona. The Vatican Advanced Technology Telescope is on Mt. Graham near Kitt Peake Observatory on Mt. Lemon, where Dick and our children visited while he was still a college student. I marveled at the confluence of my personal passions—faith and science.

After returning home, I received a letter from Father George Coyne, requesting that I serve on the Vatican Observatory Foundation Board of Directors. Don encouraged me, and I accepted. Don travels with me to Rome to participate in Vatican Observatory Board activities and sometimes to Tucson for meetings regarding the Vatican Advanced Technology Telescope at Mt. Graham.

EXPANDING OUR RESIDENCES

Don and I continued to have new adventures, including a visit to Italy that eventually resulted in our purchase of a 300-year-old farmhouse with the Nino Piccolo family in Ascoli Piceno, in the Marche region near Tuscany. Our friends Gayanne and Phillip Burns helped us clean and make repairs to get our foreign residence into shape. The morning sunrises there were marvelous, casting long shadows over the fields of olive trees and hillside vineyards. Even more spectacular was the panoramic view of the setting sun above Mount Vettore and the Apennine Mountain Range that runs the length of the Italian peninsula. The valley was peacefully quiet with only faint sounds of a distant farmer's tractor or a dog barking. It was especially quiet when it snowed. One of the grandest sights was of snow-capped mountains on a moonlit night. Stories unfolded in my mind every day as our tale of home renovation and our appreciation for the Italian culture grew.

Eventually, our children and grandchildren began visiting and loved our bit of earth located in the country shaped like a boot. Don and I built new friendships as we practiced our newly gained knowledge of the language and enjoyed invitations to see the world through different eyes and expand our horizons physically, mentally, and spiritually.

I felt a joyous closeness to God, whom I know loves all people, and I discovered that no matter where we live, a positive attitude is essential. How we weather storms, interact with friends, handle our opportunities, and face our challenges is greatly affected by our attitude—whether in the United States, in Italy, or anywhere else in the world.

Chapter 22

DARK SKIES, BRIGHT STARS

NASA'S COMMITMENT TO EDUCATOR ASTRONAUTS

As I expanded my horizons on Earth, Barbara Morgan, Christa McAuliffe's original backup teacher and friend, longed to expand hers into space. Since 1986, she had waited for her turn to fly on the shuttle. Barbara remained involved with NASA after the accident and continued to work with NASA's Education Division, speaking to audiences across the United States. While NASA worked to return the shuttle to flight readiness, Barbara returned home to McCall, Idaho, where she taught elementary school students until her selection as an educator astronaut had been announced in 1998. At that time, she had moved back to Houston to her new assignment at NASA Johnson Space Center to train with other astronaut candidates as a mission specialist.

I had cried for joy when I first learned of Barbara's selection. Finally, the long-awaited lessons from space would come to fruition. I remember calling Barbara to express my happiness and support. NASA administrator Sean O'Keefe announced publicly, "Barbara's commitment and dedication to education is an inspiration to teachers across the country. She embodies the spirit and desire of this agency to get students excited about space again, and I'm pleased that she'll be able to fulfill that mission from orbit aboard the Space Shuttle and the International Space Station."

THE *COLUMBIA* ACCIDENT

Eighty-seven space shuttle flights had flown safely since the ill-fated loss of *Challenger*, NASA's twenty-fifth shuttle mission. However, even all the advances in space travel weren't enough to prevent tragedy. Before Barbara Morgan could fly on her scheduled mission in November 2003 on STS-118, another space disaster occurred. *Columbia* STS-107, led by Commander Rick Husband, had an exceptional launch and mission success, but on February 1, 2003, during the landing sequence over Texas, the spacecraft broke apart, shattering the tranquility of our space-faring nation with the loss of another beloved crew of astronauts.

While preparing breakfast in my kitchen, I switched on CNN to watch a report from newscaster Miles O'Brien about the growth of our Challenger Learning Centers, for which he had interviewed me the day before. I also watched in eager anticipation for the return of STS-107. Miles stated he was waiting for the communications hold to end to hear the words of the commander as they readied to fly over Texas and land at Florida's Kennedy Space Center. Suddenly, *Columbia* disappeared from the radar screen.

We waited.

Silence.

Anxiously, I moved closer to the television. CNN programming became erratic and confusing. Miles appeared shaken; he interrupted with the announcement that, on its final approach over Texas, *Columbia* had suddenly disappeared. We all realized that there had been another terrible disaster in the space program. Something had happened to *Columbia* on reentry. Soon, radar images showed pieces of the orbiter strewn all across the Texas flight pattern.

Waiting in Florida on the side of the shuttle landing strip were the families of the seven astronauts, eagerly anticipating the safe return of their loved ones. They heard no sonic boom. They saw no contrails. Their hearts stopped as they stood in silence, shock, and quiet resolution. I identified with what could only be the most heart-pounding awakening to loss imaginable.

I called my children and the *Challenger* spouses. We all knew what the *Columbia* families were facing. I didn't know any of the astronauts personally, but the devastating news brought me to my knees. My phone began ringing, and Don graciously answered the calls and took notes. Finally, he said, "You have requests from dozens of reporters wanting an interview to learn what you think."

"What *I* think?" I asked, startled. "Why?"

"The nation wants to hear from you," Don said.

Feeling that this was happening to me all over again, I shook my head. Don kept urging me to take the interviews for the families and for the people of America.

"But what can I contribute?" I persisted.

"I don't know, but they are asking for you," he said calmly.

Hours went by before I collected my composure. Finally, I called the local Challenger Learning Center and suggested that they call together the news media so I could answer their questions.

That afternoon, when I stepped inside the center and walked toward our shuttle mockup reception area, I noticed a chair in position facing the crowd. Our center director, Tom Patty, escorted me to the chair. It was difficult, but I knew I had an opportunity to help the families of the *Columbia* crew by making comments and answering questions from reporters. I hoped to postpone the time when the grieving families had to face such questions.

AN UNEXPECTED IDENTITY

I waited for a question, but none came. The reporters stood reverently, wanting me to speak first. I took a deep breath and began.

"This is a terribly sad day for our nation and for the space program, and in particular for seven families whose loved ones did not return safely from space today."

I pleaded with the media to give the families space, to allow them time to recover from the shock and to collect their thoughts. I tried to explain what it's like to share the loss of a loved one so publicly with the nation.

One reporter asked, "Why now? We've called all day to ask you to speak. Why have you decided to say something?"

"Why?" I looked around the room. "Because someone needed to respond. I waited. No one came forward. I felt responsible to help the families. You've all been wonderful to help us tell our story about the Challenger Center, and I hoped I could help take some pressure off the *Columbia* families for just a little while."

Someone wanted to know how the accidents were similar or different. I thought about it briefly, then stated, "*Challenger* was destroyed on launch; *Columbia* on landing. Both missions had seven crew members." The similarities were remarkable, and I hurt more than I could say.

When I spoke with Kathie, I knew she felt the anxiety as much or even more than I did. She was reminded what it's like, from a child's point of view, to lose a parent publicly. She wanted to write a letter to the *Columbia* children, and I encouraged her, knowing it could help those children and our *Challenger* children as well.

On the following morning, Sunday, Don and I went to on Signal Mountain Presbyterian Church. As we left the church after the service, a number of people in our church family stopped to offer their support. One, Marshall Mize, offered the use of his private plane to fly us to Houston for the memorial service. On Monday morning, February 3, we met Mr. Mize at our local airport and flew with him to Houston. Astronauts and spouses waited for us at the airport and arranged for us to meet others at the NASA JSC campus for a memorial service.

FROM ONE GRIEVING CHILD TO ANOTHER

After the service, Kathie and I and other *Challenger* families met with the *Columbia* families and friends at a NASA facility at Johnson Space Center. While Kathie collected signatures from the *Challenger* children, I gathered the *Columbia* families in a small room to receive Kathie's letter. One by one, Kathie gave a letter to each older child and parent present. Then we embraced them and offered to help in any way we could. Speaking from seventeen years of missing her father, Kathie's letter began,

> Feb. 1, 2003
> To the Children of the *Columbia* crew,
>
> We, the Challenger children and all the children of public disasters, are hearing your hearts break, holding your hands, and hugging you from afar. You are not alone. Our nation mourns with you. But yours is also a personal loss that is separate from this national tragedy.
>
> We hope this letter will bring you some comfort now or in the future, when you are strong enough and old enough to read it. We want to prepare you for what's to come and help you on grief's journey. We want you to know that it will be bad—very bad—for a little while, but it will get better. . . .[1]

Tears were shed around smiles of gratefulness, and *Challenger* kids, now grown up, embraced young *Columbia* children. They bonded and took pictures It seemed like the process helped the *Challenger* children address their own grief.

We left the *Columbia* families with words of comfort and phone numbers to call when they needed us.

FACING THE AFTERMATH

As Kathie and I stepped from the building, Don called with a request for an interview with Fox newscaster Bill O'Reilly. I didn't feel up to the scrutiny, but Don, an O'Reilly fan, encouraged me to go through with it.

I participated in the interview but avoided one of Bill O'Reilly's questions about the accident. He had asked me, "Didn't the accident anger you?" Instead of talking about *anger*, I told him how our families worked to overcome our sorrow by creating opportunities for students to participate in something purposeful.

When we returned home that night, Don commented, "You didn't answer Bill O'Reilly's question. You took the high road." I fell into his arms, shaken by the day's events.

In the following days, filled with memorials and funerals, one particularly beautiful service took place at the Washington National Cathedral near my former home in Georgetown. Major General Charlie Baldwin, Air Force Chief of Chaplains, delivered the sermon, culminating with the words, "when it's darkest out, the stars shine the brightest." The message was primarily for the families of the lost *Columbia* crew, but it had significant meaning for the entire NASA family and for our nation that identifies with its pioneering efforts of the national space program.

As time moved us forward, I felt especially connected to the *Columbia* commander's wife, Evelyn Husband, sending flowers and calling regularly to listen and answer questions. I encouraged her to write her story, *A Higher Calling*, because I knew it could help her and others. I invited her and other *Columbia* families to join us at Challenger Center if they needed a creative outlet for their pain. Lani McCool, the pilot's wife, joined us at our meetings. I adored her spirit and felt an immediate kinship of empathy for her and the other spouses who shared their grief with a nation.

As we all tried to help these families deal with the aftermath of the tragedy, several of the Challenger Center board members suggested that we host a dinner event in Houston to acknowledge the *Columbia* heroes.

RED AND ROVER INSPIRE THE "COSMIC RENDEZVOUS"

A fitting tribute idea resulted from the comic strip "Red and Rover" by Brian Basset. It depicted the boy and dog sitting in a NASA box and gazing at three sets of stars that represented the three *Apollo* astronauts, the seven *Challenger* astronauts, and the seven *Columbia* astronauts who perished on their space missions. "Some stars are just meant to shine brighter than others," the little boy said, "to help light our way."

The comic provided the perfect theme for our tribute to the lost astronauts. We titled our event "Cosmic Rendezvous: An Evening among the Stars." Dayna Steele Justiz, a Houston businesswoman and wife of a NASA astronaut, offered to chair the event in October 2003. Miles O'Brien, CNN anchor, emceed the evening. Our honorary co-chairs were former president George Bush and Mrs. Barbara Bush along with twenty former astronauts and nearly fifty United Sates senators and congressmen. Our dear friend Lee Greenwood provided the patriotic entertainment.

I presented a few words about our bright stars that help light our way and expressed our appreciation to everyone in attendance, especially the *Apollo*, *Challenger*, and *Columbia* families. Joe Allen, former astronaut and chairman of the board for Challenger Center, said,

> We at Challenger Center hold the crew of *Columbia* and their families in our hearts and prayers during this difficult time, and we ask that the citizens of America and the world do the same. As people throughout the nation try to come to grips with the sad news about STS-107, we urge people to heed the words of the family members of the *Columbia* crew: "Although we grieve deeply, as do the families of *Apollo I* and *Challenger* before us, the bold exploration of space must go on. Once the root cause of this tragedy is found and corrected, the legacy of *Columbia* must carry on for the benefit of our children and yours."

THE GUIDEPOSTS FAMILY

In 2004, a friend invited me to attend a Guideposts Foundation event in Palm Springs, California. At the meeting, Ron Glosser asked if Don and I would serve on the Guideposts Cabinet for the foundation. We accepted, and several months later attended the national cabinet meeting. Two enormous blessings came from this event: a new friendship with Debbie

Macomber and the opportunity to work with Guideposts Outreach Ministry in a variety of truly meaningful ways.

During one portion of the national cabinet meeting, we each told our personal stories. After I gave my speech, a woman I didn't know walked up to me to compliment me and gave a truly loving hug. It was Debbie Macomber.

The following day, I learned all about Debbie, a famous best-selling author. Debbie's presentation was inspiring as she related her stories of early years of rejection and adversity. From that moment, Debbie Macomber and I connected on a spiritual level of mutual respect and admiration. (On the way home from the meeting, I found Debbie's books in the "Famous Author" section of an airport book shop, bought one, and immersed myself in a story that beautifully portrayed women who uproot their lives to follow their destiny. Debbie has a special gift of understanding women's relationships, and we've stayed connected ever since.)

Becoming a part of the Guideposts Outreach Ministry was the other major blessing of this event. Don joined the Guideposts committee to assist in outreach to the military, and I joined a committee of people focused on outreach to children, particularly those who were hospitalized. To help sick children, our committee created Comfort Kits that included star-shaped pillows, journals, and stickers. With the generous support of people like Sharon McKee of McKee Foods Company, we distributed these at the local hospital. Eventually, the Guideposts organization began creating beautiful kits for children that are now given out at many hospitals.

TWENTY YEARS

In 2006, we marked twenty years since the loss of our loved ones, but more important, we honored their scientific and educational legacy that lives on each day when children fly a space mission at a Challenger Learning Center. Wendy Owens, a Challenger Center board member, planned a dinner in Houston hosted by the aerospace company Lockheed Martin and our friends from the astronaut office at Johnson Space Center. It was a spectacular evening of food, music, and celebration for two decades of Challenger Center success.

Invited to speak by the Memorial Foundation at the Kennedy Space Center on January 28, 2006, I chose as the title of my speech "To Celebrate Their Legacy" (included in the appendix). Standing at a podium near the memorial in the warm Florida sunshine, I looked out to a wave of smiling faces, many who had worked alongside Dick Scobee for years. In the audience reassuring me with an encouraging smile were Don Rodgers and other *Challenger* family members. I recalled our lives at the time of the disaster, then our later shared pain with the families of the *Columbia* families. I reminded the audience of the importance of the *Challenger* legacy as carried on by the Challenger Learning Centers. I proudly introduced my son, Rich, who spoke after me.

Rich thanked the Memorial Foundation for the honor, told favorite stories about his dad, sports, flying, dinner conversations, cars, why he chose the Air Force and a career as a pilot, what it was like to fly over the Super Bowl, and why his dad continues to inspire him today as a pilot and as a father. He spoke of what the day—the honor—meant to him and thanked those who continue the NASA mission through space flight.

Though we had shared many memorials, many memories with friends and with the nation, there was something else yet undone, so in August 2006, I returned with my children to Kennedy Space Center to visit the burial site of the *Challenger* space shuttle.

SEEKING CLOSURE

In an e-mail to an official at Cape Canaveral, Rich wrote the following note about our purpose:

> My mother, sister and I have waited over 20 years for this "closure" for lack of a more meaningful term. It has been a long time and I am just now able to come to grips with the loss. As you may or may not know, I am a Colonel in the Air Force now and have been a fighter pilot my entire career. I have certain goals for this trip, and I don't think I'm doing a good job of communicating them.
>
> First, and most importantly, we want to see, touch, and feel the actual *Challenger*. From our phone conversation it sounds as if it is sealed in a silo. Could you please explain? Is it a silo that can never be opened or one that no one wants to open?
>
> We wanted to talk with someone who is familiar with the accident and what happened. It has been so long, there may not be anyone available. I've been through the techno babble of the Rogers report, and we'd like to talk with someone that speaks pilot and can explain any questions we might have.
>
> I am aware there may be pictures that can be viewed and not released to the public. We would like to see them.

I am flying my F-16 into Patrick Air Force Base on Thursday doing some instrument training . . . are there any restrictions in flying over launch pad 39A? This is the culmination of my Air Force flying career and I want to visit my father's launch site once before I hang up my spurs.

Thank you for your patience and your time.

Warm regards,
Rich, June, and Kathie

The three of us rendezvoused at Kennedy Space Center. We met an official who graciously took us on a tour to visit the silo. After the opportunity to walk around the silo, Kathie recorded her memory of the experience:

> Very few people know about an abandoned corner of land at the Kennedy Space Center where the remains of the *Challenger* rest. There is no marble memorial. No signage. No faded silk flowers. There are old, sealed underground missile silos containing crated, buried, crippled pieces of a space vehicle.
>
> When we traveled to the Cape for emotional closure, we were met by a very nice engineer or manager who was assigned to us for this, our last mission, to absorb the reality of the *Challenger* explosion that rocked our world. Indeed, it felt like this man was the only person left in that world who knew of the lonely graveyard and its contents.
>
> We thought we could see the orbiter pieces—touch them—finally say good-bye to our Dad and his final mission. . . . For us, the journey had gone on too long. Far too long.
>
> When we reached the site, it was clear the silo was completely sealed. The only way to access the contents would be to enlist the aid of a crane to remove the lid. This was something NASA was unwilling to do.
>
> They had already conducted their research and cataloged the remains. They had already pieced together the orbiter's wreckage found at sea. They crated up the evidence. They buried it. They were done. They had moved on.
>
> So we sat together on the sealed lid of the concrete silo among the weeds of a desolate, forgotten spot and said good-bye one last time. And we began to move on, too.

Rich had a final salute to his father and the beloved *Challenger* space shuttle. To Rich, the orbiter was more than scrap metal buried in a silo. It represented many successful missions to space, including his dad's first flight in 1984. With permission, Rich flew over the Cape. But instead of feeling remorse, he told me he felt pride for his dad and also, as he flew over space shuttle *Discovery* poised for launch, for the entire astronaut corps who continue the mission.

All three of us agreed that there's no way we could ever say enough about Dick Scobee. I can never explain how amazing he was. Someone so incredible never truly leaves. He's still part of our family, and we talk about him all the time.

Of course, we feel that Don, my husband and their stepfather, is just as amazing. For many years, he has filled the gap of love Dick left, adoring me, the children, and the grandchildren. On Father's Day, Kathie and Rich always think of Dick, but the memories don't make them sad anymore. They appreciate Don's love and his example of someone to emulate.

Note

1. For the complete text of Kathie's letter, please see the appendix.

Chapter 23

STAR QUEST

THE EDUCATOR ASTRONAUT

After the *Columbia* tragedy, Barbara Morgan's trip to space had been post-poned. Loyal to her commitment and dedicated to her mission of space education, Barbara kept pursuing her dream, and her flight was finally rescheduled for August 2007. As the mission plans unfolded, our Challenger Center staff applied to NASA to be one of several downlink opportunities, which meant Challenger Learning Centers would have a direct video connect to Barbara while she was in space.

We were ecstatic to be chosen for the NASA opportunity and began planning to reach thousands of students with Barbara's downlink from the International Space Station. After raising sufficient funds with the help of generous supporters, we were ready to get started.[1]

BARBARA MORGAN HEADS INTO SPACE

The launch of STS-118 at Kennedy Space Center was my first to witness since I watched the *Challenger* accident. In spite of my concern for the safety of the flight, I was excited for Barbara's long-awaited mission to space. Along with mission specialist Barbara, *Endeavor*'s mission crew included Commander Scott J. Kelly, pilot Charles O. Hobaugh, veteran astronauts Richard Mastracchio and Dave Williams of the Canadian Space Agency, and Tracy Caldwell and Benjamin Alvin Drew as additional mission specialists.

On Tuesday, August 7, 2007, Don supported me as I gave interviews to *Fox Morning News* and the *Fox & Friends* program. Other reporters also wanted to know my reaction to Barbara's flight. Except for Barbara, her fellow crew members, and their families, I could not imagine that anyone was happier and more anxious for the launch than I was.

In an interview before the flight, Barbara told reporters, "I am going up doing the job of an astronaut, the work of an astronaut, but I'm going up with a teacher's eyes, ears, heart and mind. And so I look very much forward to doing that with an open mind and being able to come back and . . . translate that into how we can best provide wonderful opportunities for our colleagues and our students." She also said, "Christa was and is and always will be a great representative of the teaching profession, and we are really, really proud of her. She was, is, and always will be our Teacher in Space."

When reporters asked for my opinion, I said, "Barbara is persistent, determined, and she still has that youthful bubble and spark of enthusiasm that she had as a young woman when she first was training alongside Christa McAuliffe. She is representing educators everywhere, showing that it's a truly honorable profession. She's [showing people that] persistence pays off and saying to kids, reach for the stars, keep plugging away at those dreams. What a message she's getting out to the general public in addition to teachers and kids."

As we moved away from the media, Don escorted me to the VIP Visitor Complex to view the shuttle launch. It was a rather new facility created by NASA especially for families and close friends of the crew members aboard the shuttle. After the final countdown began, visitors and NASA employees walked onto a balcony to view the launch. Don held me as I prayed, then gasped while the shuttle engines ignited and *Endeavor* slowly climbed upward, clearing the tower and roaring into the sky with more than seven million pounds of thrust from a steady stream of bright orange fire.

I shouted to Don over the crackling thunderous blast, "She's made it!" I watched the shuttle climb higher until it rotated and glistened like a heavenly star. I prayed that God would hold Barbara safely in his arms; I cried tears of happiness and heaved a great sigh that shed decades of tension.

STS-118 DOWNLINK

Within days, I traveled to participate in the downlink activities at Challenger Center headquarters in Alexandria, Virginia.

Challenger Center had been chosen in NASA's open competition to have a twenty-minute live question-and-answer downlink session with mission specialists Barbara Morgan and Benjamin Drew on August 16. On the day before, an open house celebrating the mission was held, including student activities, NASA exhibits and astronauts, and an interactive astronaut webcast with Challenger Learning Centers. (See www.challenger.org for links to the KZO Networks broadcasts of both the astronaut webcasts and the STS-118 downlink.)

Twenty-one years after we originally planned to have lessons from a teacher in space, the day finally arrived for our dream to come true. I could barely contain my excitement as a voice from Mission Control said, "*Endeavour* ISS, this is Houston. Are you ready for the event?"

"We are ready," came the voice of Barbara Morgan.

As I stood with children from all over the country, I was euphoric and greeted Barbara with words I'd practiced for this historic moment, "Congratulations, *Endeavor* crew. Barbara, we've been standing by at the Challenger Center waiting for your call for twenty-one years."

It was time for students and the media to take part in a live downlink with STS-118 astronauts. Through the wonders of technology, we could see and hear the astronauts, and they could see and hear us.

"We are absolutely thrilled to be talking with you," Barbara said to the assembled students at the Challenger Center and countless others participating via Challenger Center Internet. "Good morning from ISS."

Students from across the country asked questions ranging from "How do you brush your teeth in space?" to "How will the experiments on the ISS help continue and further the mission to Mars?"

Not only did Barbara and Benjamin answer the students' questions, but they also demonstrated some of their answers. "How and where do you sleep?" asked James from Florida, so Benjamin demonstrated getting into a vertical sleeping bag and zipping it up so it would hold him in place in zero gravity. Barbara washed her face with a spray of soapy water, demonstrating how astronauts keep clean in flight.

Sarah from Indiana asked, "If an Olympic-sized swimming pool could be built in space, would you be able to swim faster on Earth or in space, and where would you burn the most calories?" To demonstrate why a celestial swimming pool is not practical, Barbara and Benjamin pulled out a bag of water and squeezed enough out from a straw to form a big ball that floated in front of their faces. When Barbara put out her hand, the ball hit it, and then broke into floating pieces. Benjamin opened his mouth like a giant fish and took in a bite-sized glob of water.

"What is your favorite space food?" asked Jessica from Illinois. Holding up a bag of M&Ms, Barbara tossed a red and then a blue one to float into Benjamin's mouth. Then she admonished, "Kids, even in space you have to brush your teeth. In space there are no sinks, so astronauts have to spit into a small towel."

Asked how it felt to enter space for the first time, Barbara declared it as "absolutely wonderful, and I imagine that going into space for the hundredth time is also wonderful."

When Maddy from Washington, DC, asked whether Barbara had a special teacher or mentor, Barbara replied, "That was the *Challenger* crew. They were my teachers, and I believe they're teaching us today, too."

When I returned home to Chattanooga, Don greeted me enthusiastically knowing full well how important the previous days had been to me and to a nation of teachers and their students. Our local media had covered the event, and I was thrilled to read the account (see the appendix).

THE CHALLENGER PRESIDENT GEORGE H. W. BUSH AWARD

Within weeks, Challenger Center representatives traveled to the Bush Presidential Library at Texas A&M University so the former president could present the special award for leadership in continuing the Challenger Center mission and fulfilling the dream of the families of the 51-L crew.

After greeting the Bushes, my dear friends, I introduced the former president to our chairman Bill Readdy and to Jon Clark, who was with us to represent the *Columbia* crew. Then Mr. Bush lovingly presented the Challenger Center President George H. W. Bush Award to Barbara Morgan and said, "With this award, we commemorate your historic flight, honor your dedication to the teaching profession, and acknowledge your commitment to the mission of the Challenger Center."

In 1995, President Bush and First Lady Barbara Bush had been the first recipients of this award, which was subsequently named after the former president. It is the Challenger Center's highest honor. Only those who display exceptional leadership and who contribute significantly to the center's mission receive this honor.

Note

1. Without the generous support from sponsors for the Challenger Center downlink and broadcast across the nation to all our Challenger Centers, we would not have been able to connect Barbara and her lessons to our teachers and students. The sponsors were Challenger Center and my dear friends who shared in our joy to see a teacher fly in space and demonstrate answers to questions from students across the country. Sponsors included Republic Parking with Jim Berry, Lockheed Martin Co. Foundations, and families from SunTrust Bank and CANDL, Ellsworth McKee, Jack Maxwell, Hardy Caldwell, Fran and Christie Simmons, Bobbye Harris, Steve and Lou Brown, Colleen Phillips, Steve McAuliffe, and the Scobee-Rodgers families. KZO Networks provided webcasting services, and Valpak advertised the event across the country.

Chapter 24

COSMIC ADVENTURES

RICHARD GARRIOTT, A NEW KIND OF ASTRONAUT

Richard Garriott, who helped us develop the original concept for the simulated experiences at Challenger Learning Centers, is the kind of resourceful hero Americans admire. He worked hard to carve out his path to success and is now an astronaut extraordinaire, an exemplary citizen of both the United States and England, and a self-made millionaire. As a friend, I am happy for him. As his former teacher, I am proud of him.

Richard was the keynote speaker for the 2009 Annual Challenger Center Conference. Using slides of his experiences aboard the International Space Station (ISS), he explained why it was so important for him to travel to space, conduct scientific experiments, and organize learning activities for students. He was a natural, a new breed of explorer, a space pioneer, and a hero.

Listening to his story, the teachers and flight directors were spell bound. Lucy Hawking, famous author and daughter of the well-known scientist Steve Hawking, said of British-born Garriott, later wrote to me in an e-mail, "[There's] a new type of space entrepreneur emerging, stylish, swash-buckling space adventurers who may, ultimately, prove to be the greatest inspiration of all to the younger generation. One man who will probably turn out to be responsible for an unprecedented number of astronaut applications and careers in the space industry in the future is private astronaut Richard Garriott."

A LIFELONG DREAM OF SPACE ADVENTURE

Born to Helen and Owen Garriott, Richard learned from his father's space journeys on the 1973 Skylab 3 mission and on the 1983 Spacelab-1 mission. At the conference, he shared how he had grown up surrounded by astronaut families just around the corner from NASA's Johnson Space Center. "I grew up believing that everyone could travel into space," he said. Unfortunately for the young Garriott, a NASA doctor told him he would never be selected as an astronaut because of his poor eyesight. "I thought to myself that if I could not be a member of 'their' space program, I would just have to go and make my own," Richard explained. "And I did!"

Twenty-five years earlier, Dick and I had lived in Houston, where Dick trained to be an astronaut and worked with Richard's father, Owen. As a youngster, Richard was curious, creative, verbally precocious, and already an innovator. He and his mother, Helen, invited me to their home to join a group of students and learn how to play *Dungeons and Dragons*, a fantasy role-playing game. Richard's keen grasp of the intricacies of games helped him as he went to college and immersed himself into creating computer games.

By the time, Dick Scobee had flown in space, Richard was already a multi-millionaire. After the *Challenger* accident, Richard willingly helped me figure out how to link the computer stations at our Challenger Learning Centers, eventually working with others such as artist Bob McCall to finalize the simulation development. Still, Richard never lost his dream to go to space himself.

TICKET TO SPACE

With part of his fortune, Richard invested heavily in space-related enterprises, hoping it would lead to an opportunity to fly with the Russian space program. Eventually, his space flight was set for October 2008 to the International Space Station (ISS). Most admirable about Richard is that he wanted to use his space time for scientific experiments. While on the ISS, he undertook research into protein crystal growth and created an opportunity to make a difference for students, continuing his initial work with Challenger Center.

Richard also spent time answering questions from students who represented the US and UK Challenger Learning Centers—both while he was aboard ISS and after his return. The former student even got to teach his former teacher. I asked Richard to tell me what he considered the three most memorable or most challenging kids' questions. He was pleased with all the questions and enjoyed responding to each child, but at my request, he selected three for me to share.

Question 1. Is it hot or cold in space?
Answer: Both! Space being a vacuum has no direct temperature, but if you are in the sun, thermal loading will make you burn up; if you are in the shade you will slowly radiate your heat away and freeze.

Question 2. If you strike a match in space, will it ignite?
Answer: (a) Fire requires heat, fuel, and an oxidizer. The tip of a match has both fuel and oxidizer; striking it creates heat, so the tip will burn, but when the oxidizer on the tip is used up, it will go out! (b) Inside the space-ship, where there is oxygen, the match will also go out after the tip burns because of the lack of convection to carry away the CO^2 and bring in fresh O^2.

Question 3. Can you burp in space?
Answer: Standing on your head on Earth makes it hard to burp. In space you have a similar problem, but after a couple days your body figures it out!

(For more about Richard's amazing experiences as a private astronaut, see his website at www.richardgarriott.com.)

While there, he said,

In this day and age when we clearly see and feel the finite scale of the Earth and how we are all interdependent, space more than ever represents the great wonder that lies beyond. I think this is true for our generation even more than the previous. The previous generation was trying to prove the capabilities of a country and an ideology. Our generation is looking to see how space can benefit humanity here on Earth and how we grow as a species to live beyond the confines of the Earth where our species was born, but will surely not remain here alone.

His former teacher couldn't agree with him more.

Chapter 25

TO THE STARS WITH MOON DUST

PER ASPERA AD ASTRA

The phrase *per aspera ad astra* is Latin, meaning "through difficulties to the stars." In other words, we must be prepared to go through hard work and sacrifice to reach our highest goals. With origins as early as Seneca and Virgil, the words are used in varied forms as part of the motto of many organizations, such as the British Royal Air Force, who adopted the words that translate to "Through adversity to the stars."

The translation from Latin reminds me of my own moments of joy and of profound sadness. I'm acquainted with life-altering sorrow, yet I work hard at focusing on the positive to shape the possibilities of tomorrow, always searching for the silver lining beyond the dark clouds.

It was rather symbolic for me when Richard, prior to his time on ISS, invited me to Austin, Texas, to fly weightless in a Zero-G aircraft.

Recalling that every time a reporter asks me if I would fly in space, I always answer without modesty, "When NASA needs a grandmother in space, sign me up," I felt hesitant at first. But then I realized it was my opportunity for a similar experience, so I quickly agreed.

Eventually, the day arrived when I flew to Austin, met other participants at the airport, and listened to a brief introduction about the safety of the flight, procedures, and instructions for donning my own spaceflight suit. Before I knew it, we had entered the plane, strapped ourselves in passenger seats at the back of the cargo area, and took off. The aircraft climbed to a pre-set altitude where we were told to remove our seatbelts and lie on the carpeted floor. The plan was for the pilot to fly parabolic maneuvers that began small, allowing us the experience of weightlessness similar to that in low Earth orbit, then increased so we could get a sensation more like that on the moon and then Mars. Due to the exacting procedures, no one felt nauseated or got sick.

The adventure was marvelous. I turned somersaults, sipped balls of water from the air, flew like Superwoman across the cargo bay, and squealed like a kid. I didn't want the experience to end, but the unmistakable pleasure of the opportunity gave me an unexpected delight. I learned firsthand the joy Dick had expressed the night after he returned from his first flight in space.

AWARDS FOR CHALLENGER CENTER

As I achieved a personal goal of flying in zero gravity, Challenger Center continued to meet the goal of inspiring children to use math and science to explore space. Knowledge of the center's work spread around the world. It was a tremendous honor to receive the 2007 National Space Club Educator Award at their 50th Anniversary Goddard Memorial Dinner in Washington, DC. This award recognizes significant achievements in space science and enterprise. Also in 2007, I became the first woman to be the featured speaker at the annual banquet of the Southern Surgical Association during the meeting held in Hot Springs, Virginia. In October 2008, I spoke at the 59th International Astronautical Congress in Glasgow, Scotland. The conference theme, "From Imagination to Reality," aptly spotlighted the Challenger Center story. I also traveled to Seoul, South Korea, to meet the family who built a Challenger Center and observatory in a vast facility surrounded by luscious groves of trees and gardens.

In Washington, DC, in May 2009, the American Institute of Aeronautics and Astronautics (AIAA) awarded its Foundation Award for Excellence to the Challenger Center for Space Science Education. The ceremony took place at the Aerospace Spotlight Awards Gala held at the Ronald Reagan Building and International Trade Center. As founding chairperson, I accepted the award for the center. Given annually to recognize unique contributions and extraordinary accomplishments by organizations or individuals, the award honored the center's "two decades of inspiration and fostering interest in careers in science, technology, engineering, and mathematics."

Personally, I received the unexpected honor of being nominated for the Horatio Alger Association of Distinguished Americans. It bears the name of the renowned author Horatio Alger, Jr., whose tales of overcoming adversity through unyielding perseverance and basic moral principles captivated the public in the late nineteenth century. His name endures as a symbol of our most basic and valued belief: that, here in America, any man or woman can triumph over adversity and achieve success despite humble beginnings. The association inducts new members every year who, regardless of their under-privileged backgrounds, have become leaders and high achievers.

Though I remained among the top fifteen nominees, only ten are announced as winners. I have tremendous respect for association member Terry Giles, who nominated me. It's a great honor to have been nominated and to have my name listed alongside other truly great Americans.

STAR CHALLENGERS SERIES

Another opportunity to promote Challenger Center and inspire the imagination of children arose when well-known author Kevin Anderson contacted me about co-writing a science-fiction story about teenagers and their adventures involving space. Kevin and his wife and fellow author Rebecca Moesta were known for their award-winning and *New York Times* best-selling *Star Wars: Young Jedi Knights* series, their work on *Star Trek*, and their *Crystal Doors* series. Together, they had sold more than 20 million copies around the world in more than thirty languages.

Intrigued by the idea, I met with the couple, and together we began devising a plot for a story about teens in space. In the following months, Rebecca and I developed the characters, but publishers weren't interested, citing the fact that science fiction isn't popular—only science fantasy. We were told to add magic, dragons, and wizardry, but we held strong to our original convictions.

Finally, Catalyst Game Labs was interested. With exciting stories and interesting, relatable characters—including young men and women of various backgrounds—we hope to foster a generation of thinkers who choose careers in science and technology.

In the first book, *Star Challengers: Moonbase Crisis*, four students are chosen to attend an event at a nearby Challenger Learning Center run by the mysterious Commander Zota. Soon, they are swept into a wild adventure that will take them across time and space, for Zota is from a bleak, desperate future where Earth faces a terrible crisis with no innovators to help solve it.

Zota has traveled back in time to help create a generation that will be prepared for the challenges of the future. (More volumes are planned, and information about the series is available at www.starchallengers.com.) The story emphasizes to young readers that their educational interests and career choices can have an enormous impact on the future and may even save the world.

First published in fall 2010, the series of young adult books is another of my fulfilled dreams for helping young people realize they can overcome any circumstances and reach for the stars. We live in a society increasingly governed by science and technology, and yet fewer young people want to go into science. No doubt, Challenger Center is needed more than ever to seek solutions to inspire and motivate our youth to study these fields. *Per aspera ad astra*—through adversity to the stars!

Chapter 26

GOLDEN HORIZONS

NEW HORIZONS

What beckons us toward new horizons? Is it the ever-changing seasons of life? Or is the hope that after the storm, we can see the sun's powerful effect to help us discover the purpose in our lives? Is it a continuing need to grow, to change, or to affect the world around us as a part of God's plan? Or is it simply that life is what happens on our way to fulfill our dreams?

Dick Scobee and I had great plans. We thought we had discovered our life's purpose. We were soul mates traveling fast down that path. Suddenly, it ended. But my life didn't. At first, I dangled at the dead end stop as he climbed on heavenward. I wanted to go with him, but it was not my turn. I waited at a standstill, grieving and searching for my way. After some time, I could see the horizon and realized my path had veered off and taken a turn at an intersection. What I learned from others is that a tragic event in life such as illness, divorce, or even death does not determine our destiny. Instead, how we react—how we manage the bumps, hurdle the roadblocks, and handle the dead ends—determines our way.

My strength came when I accepted my problems as challenges and resolved to overcome them. Within each struggle was the opportunity for growth. Sometimes I succeeded, and other times I failed, but when I was successful, it was because I relied on my lifelong "ABC" guide of optimism and faith that I discovered in my childhood. It was the day as a child, eating those tangerines and reading the Peale book *Power of Positive Thinking*, that

I learned about the importance of attitude and the courage to believe in my dreams and the power of prayer.

In time, I learned that sometimes God answers prayers through people. I've been blessed with a wonderful family and dear friends who have helped me over life's hurdles, led the Challenger Center organization, and helped rescue us from a spiraling downward path that resulted partly from the slowing economy. In particular, former astronaut and board member Bill Readdy took on the leadership role of Challenger Center chairman, helping the organization refocus on its mission and improve the fiscal situation of 2009. Dan Barstow, who had extensive experience in the fields of Earth and Space Science Education and advanced skills in utilizing the latest technology, was selected as president, and he worked to review the processes to get Challenger Center back on a successful path.

Additionally, chair-elect Scott Parazynski, also a former astronaut, is dedicated to the mission and works well with Dan to keep the organization sustainable through hard work, hands-on leadership, and facing the facts. Scott organized a retreat for the Challenger Center board members and shared his history and philosophy with us: "My own passion for space began at age five, growing up in the shadow of *Apollo* and my father's work on the program. Although our present-day space shuttle is about to retire and the path ahead for human spaceflight is less than clear, Americans will continue to fly into space on one vehicle or another—and a new generation of young explorers is watching and waiting for their turn." He quoted our twenty-sixth president, Theodore Roosevelt, who said, "Far better is it to dare mighty things, to win glorious triumphs, even though checkered by failure . . . than to rank with those poor spirits who neither enjoy nor suffer much, because they live in a gray twilight that knows not victory nor defeat."

THE FUTURE OF CHALLENGER CENTER

When I asked Challenger Center president Dan Barstow for insight into his personal vision and mission for Challenger Center, he said,

> We are at a pivotal point, for our nation and for Challenger Learning Centers. Our nation needs to inspire and engage young people in science, technology, engineering, and math. Our nation's excellent teachers do a fine job providing content knowledge, but often don't have the time or opportunity to add the excitement and passion that make science come alive. That's where Challenger Learning Centers make their great contribution to our nation. As William Butler Yeats said, "Education is not the

filling of a bucket, but the lighting of a fire." Challenger Learning Centers light fires!

Over the next decade and beyond, with Dan Barstow's leadership, we want to find new ways to extend and deepen the passion for learning. Challenger Learning Centers will change with the times. The space program is undergoing a major shift, with the end of the space shuttle program in 2011 (as NASA retires this formidable work horse), the emerging role of private space travel, and new opportunities for young people to participate in space exploration.

Dan is leading Challenger Center to embrace and integrate these new ideas, including helping students work directly with scientists and engineers to design, launch, and run their own space experiments. He is assessing the opportunities provided by telecommunications technologies, which can allow students to participate in CLC missions from school and home computers. Also, social media can help build large communities of CLC alumni and others who are interested in space exploration.

As an organization, we are also planning to take students on a mission into the ocean, adapting the simulator to work like a submarine, with the associated story line, learning experiences, and animations of a Captain Nemo-like adventure on a research vessel. Another idea involves transforming the simulator into a tiny robot that goes through the human body on a biology mission, providing medical attention to an astronaut serving with a crew on another planet or a remote asteroid. These ideas build on the same core vision—that we engage and inspire young people in science, technology, engineering, and math, and that we do this most powerfully by immersing them in the real (or realistically simulated) world of astronauts and other explorers, scientists, and engineers.

With new and strong leadership in place to focus on our core mission, I've learned that we must be ever vigilant to the mission that was created with passion and continues with love. Across our current network of fifty Challenger Learning Centers, we have an incredible team of educators and scientists with talent and passion for our mission. We will grow and thrive over the next decade and beyond, and this group of talented people will keep our mission alive. We do this for our nation and for the world. We, along with the 4 million students who have participated in Challenger Learning Center missions, will help our nation move forward.

WITNESSING THE RESULTS

Seeing the organization grow to reach more teachers and their students is rewarding, but making a difference in the lives of individual students is gratifying beyond words. Every year we hear from the students, their parents, or their teachers and leaders in Challenger Learning Center communities everywhere. They tell us about their opportunities and thank us for making a difference in their lives.

One adult from New Jersey described a breakthrough that came for her when she attended the Buehler Challenger Learning Center as a student. Now a NASA Engineer, Meg Meehan shared with the center director, Kathie Klein, that she would be a mission controller for STS-125. "I can't tell you how much it means to me to be considered one of the center's success stories," Meg wrote to Kathie in 2008. "I'm where I am today, no doubt, as a direct result of the experiences and opportunities I had at the Challenger Center. I don't think I could ever adequately thank you all; just know that when I'm sitting in Mission Control next summer, I'll be wearing a Challenger pin proudly and thinking of you all." (For more CLC success stories, please see the appendix.)

It is just as rewarding to hear the teachers, flight directors, and educators speak of the opportunities the Challenger Learning Centers provide. Lead flight director Becky Manis said, "Our Challenger Learning Center has become an annual education destination for thousands of students and hundreds of educators, parents, and visitors. We're proud to welcome students of all abilities and economic levels. We smile when we hear a student gasp as she walks in the building for the first time; but we swell with emotion when we so often hear a student say on her way out, 'This is the best day of my life!'"

Carol O'Leary, former director of the Colorado Challenger Center said, "Colorado Springs is heavily invested in space programs. When we employ space mission simulations to attract students into mathematics, science, and technology careers, we are filling the pipeline for very lucrative and exciting careers right here in our community. Industry and others are beginning to understand that value."

Superintendent Daniel A. Domenech said, "Fairfax County Public Schools is pleased to endorse the Challenger Learning Center of Greater Washington. At a time in our history when the fields of mathematics, science, and technology are critical to our country's future, programs such as those provided by the Challenger Learning Center light a spark of enthusiasm in our school children."

Leaders in our nation believe in the opportunity as well, as these supporters indicate:

• What Challenger Center has done with respect to educating America's youth is truly commendable. I salute you. (General Colin Powell, USA (ret) Secretary of State)

• I know one good answer to the questions people are asking about American education, American scientific leadership, and American students' motivation to learn. That answer is Challenger Center. (Buzz Aldrin, astronaut, *Gemini* 12 and *Apollo* 11)

• The mission of Challenger Center is to spark in our young people an interest—and a joy—in science. A spark that can change their lives—and help make American enterprise the envy of the world. (President George H. W. Bush)

Chapter 27

STARRY NIGHTS

OUR VAST UNIVERSE

The year 2009 was the International Year of Astronomy, celebrating the 400th anniversary of Italian astronomer Galileo's first use of a telescope to observe the cosmos. The celebration commemorated Galileo's observations with the telescope that he constructed, which forever changed our view of the universe and ourselves.

Anyone who has looked to the night sky or studied the stars knows the universe is vast and intimidating. How do our scientists, teachers, and students at Challenger Center even begin to fathom objects and distances that dwarf anything we've ever experienced? Our galaxy possesses billions of stars; our sun is 93,000,000 miles away. To many students, these numbers are irrelevant. On occasion, I've tried to explain using an idea I found in Chet Raymo's *365 Starry Nights*.[1] He uses an analogy about the galaxy using boxes of salt.

When given an opportunity, I love to speak to the students at the CLCs about the wonders of space. I ask, "If you want to make a realistic model from salt where every salt grain represents a star and they are at properly scaled distances, how many cylinder salt boxes do you think it would take to represent all our stars?" I flip open the top of a box of salt and pour the contents on a sheet of newspaper, allowing the children to guess how much salt. They often respond with a handful of salt, or measurement like 1/2 cup, or a pinch, and then a bright youngster says, "All that salt."

How much? According to Chet Raymo, it would take 10,000 boxes of salt spread over the area of the orbit of the moon.

SCIENCE AND RELIGION: EXPLORING GOD'S UNIVERSE

Using the technology created by man, we can lift up our eyes to the heavens in order to rediscover our place in the universe. After contemplating the sky long ago, the psalmist wrote, "When I see your heavens, the work of your fingers, the moon and the stars which you set in place, what is man that you should be mindful of him, or the son of man, that you should care for him?" (Ps 8:4-5). The more I learn about the universe and contemplate its creation, the more I marvel at the Creator and of the love that motivates God.

I've come to believe that God is bigger than the universe, even bigger than all the stars in our galaxy we see at night and the ones we can't see. God created it all. This creation reminds us of God's power and greatness, and also that God is everywhere. God is a spiritual being not limited to one place or one time. God is everywhere, here and already there. I believe God invites us to discover and to learn about our place among the stars to help us begin to grasp the mighty universe.

The Vatican Observatory near Rome, Italy, serves as a bridge over time and space, drawing together a rich history and a current engagement in the search for meaning. The most outstanding astrophysics and astronomy students from around the world come annually to appreciate the beauty of God's universe. Members of the Vatican Observatory come from four continents, speak nine languages, and work in almost every field of modern astronomy: Big Bang cosmology and string theory, galaxy formation and evolution, stellar spectroscopy, meteors and meteorites. Our Vatican Observatory and headquarters is located at the Pope's summer palace in Castel Gandolfo, in the hills south of Rome. Every other year, we train aspiring astronomers by sponsoring a four-week summer school. The students, most from developing countries, are given an intensive course in some aspect of astrophysics from world experts while developing friendships that will last a lifetime.

Astronomy has had a huge impact on our culture. For those of us who are believers, astronomy opens our minds and hearts to the Creator. Perhaps even for non-believes, astronomy helps them to realize the magnificence and beauty of the universe of which we are a small part.

A meeting held in Florence, Italy, in May 2009 was one of the major events among many to celebrate the International Year of Astronomy. Speaking at the event was Father George Coyne, SJ and president of the Vatican Observatory Foundation for which I serve as a board member. He provided excerpts from letters of Galileo, who expressed gratefulness to God that he alone was the first to observe the marvelous things that had been hidden in all ages past. While Galileo could not bring the stars to Earth, with his telescope he took the Earth toward the stars and spent the rest of his life drawing the significance of those discoveries.

Another event took place at the Vatican in October 2009. Pope Benedict XVI addressed an international group of renowned astronomers and the Vatican Observatory members. He said,

> This celebration, which marks the four hundredth anniversary of Galileo Galilei's first observations of the heavens by telescope, invites us to consider the immense progress of scientific knowledge in the modern age and, in a particular way, to turn our gaze anew to the heavens in a spirit of wonder, contemplation and commitment to the pursuit of truth, wherever it is to be found.
>
> I am particularly grateful to the staff of the Observatory for their efforts to promote research, educational opportunities, and dialogue between the Church and the world of science. . . .
>
> Who can deny that responsibility for the future of humanity, and indeed respect for nature and the world around us, demand—today as much as ever—the careful observation, critical judgment, patience, and discipline which are essential to the modern scientific method? At the same time, the great scientists of the age of discovery remind us also that true knowledge is always directed to wisdom, and, rather than restricting the eyes of the mind, it invites us to lift our gaze to the higher realm of the spirit. . . .

FROM GALILEO TO HUBBLE SPACE TELESCOPE

From the Vatican Observatory observations established in 1891 by Pope Leo XIII to the spectacular views to the cosmos provided by the Hubble Space Telescope launched by NASA astronauts aboard STS-31 in April 1990, the human race has learned that Earth is not at the center of the universe. Instead, the universe is composed of billions of galaxies, each of them with myriad stars and planets. Over the years, Hubble imagery has amazed people around the world, helped scientists determine the process of how planets are born, and assisted in the rewriting of astronomy textbooks.

STS-151 made a final journey to the Hubble space telescope before the shuttle fleet's scheduled retirement in 2011. Like the *Challenger* crew, the STS-151 crew members received instruction in cinematography and carried cameras on board. Their photos and others like them help astronomers peer into the farthest reaches of space and catch fantastic glimpses into the universe's distant past. The crew went up in May 2009, and with their Hubble Space Telescope captured the earliest image yet of the universe—just 600 million years after the "Big Bang," when the universe was merely a toddler. With the telescope, they zoomed past the star Sirius, a mere 50 trillion miles from Earth, to gaseous clouds in the Orion Nebula where stars are being born. They glided on outside the Milky Way to Andromeda, our closest galactic neighbor, and visited the Virgo Cluster, a heavy-traffic area containing a family of 2,000 galaxies. With their photography, NASA astronomers tell us the imagery is humbling and heavenly enough to make one feel insignificant and yet exhilarated at the same time.

Since my childhood when I wished upon the first star and looked to the heavens to find comfort in God's presence, I have seen beauty, elegance, and joy in the cosmos. If God is love, then love is the master key that opens the door to discovery, and exploring space allows us to quench our thirst for knowledge and belonging. It's a glorious calling to look beyond the confines of our limited life span and beyond our limited location in the cosmos. It's our destiny. God is bigger than the universe, and science and faith are not only compatible, but wondrously complementary.

God's mighty universe and wonders are almost incomprehensible to me, but just as incredible as the wonders of space is God's creation of a single human placed upon this planet orbiting around our star, the sun. Dick Scobee grew from the three-year old who wheeled around on his pedal airplane to the teenager who built airplane models, sketched and painted them. As an adult, he tested high performance aircraft and piloted the space shuttle to the heavens. He loved his work. It was his calling. Though his life was cut short, his legacy and that of his beloved *Challenger* crew live on every day that a child experiences the wonder and joy of continuing their mission, or a teacher inspires students to creative pursuits.

Former astronaut and senator John Glenn said that "America's space program has always been about discovery—discovery of our universe, our world, and most importantly, ourselves." Everything worthwhile starts with a creative idea. Passion for an idea is the most powerful creative force in the world. It was an idea that brought the technology forward to observe the heavens or to fly the rockets to reach them. And it was just an idea emblazoned in the hearts of the *Challenger* families to fulfill our loved ones' dreams. Their legacy lives; their love is everywhere; their Truth is forever as witnessed each time a silver lining peers from behind a cloud or a child reaches to touch a star.

> *And we know that in all things God works for the good of those*
> *who love him, who have been called according to his purpose.*
> *(Romans 8:28)*

Note

1. Chet Raymo, *365 Starry Nights* (Englewood Cliffs NJ: Prentice-Hall, 1982) 199.

Challenger Center founding directors: Chuck Resnik, Lorna Onizuka, Jane Smith, Cheryl McNair, Steve McAuliffe, June Scobee, and Marcia Jarvis. 1989.

The *Challenger* families with Challenger Center President Doug King and his staff at the 5th year anniversary, Washington, DC 1991.

June and *Challenger* spouses pose with George and Barbara Bush
at the tenth anniversary event in Washington, DC.

The eleven Floozies. Front row from left to right: Trish Foy, Fran Simmons, Helen
Exum, Lisa Frost. Back row: Sandra Longer, Ruth Obear, June Scobee Rodgers,
Nell Mohney, Kathleen Nielson, Carol Mutter, Gayanne Burns.

At Fort Huachuca, Quarters 1, Girl Scouts sell the first cookies
to the commander, Lt. Gen. Rodgers, and June. 1989.

Our family. Back row: Eric, Anne and Margaret Rodgers, Kathie and Scott
Fulgham, Dexter Scobee, the bride and groom Alene and Rich Scobee, Emily
and Justin Krause. Front row: Charlotte and Jack Rodgers, Jilly Fulgham,
Andrew and Cristi Scobee and Amma June and Papa Don.

June and Don on the Kawasaki Sport
Touring motorcycle, getting ready
for their "four corners" motorcycle
adventure in 1991.

June with Vance Ablott, Challenger Center president, and Bob McCall at
Phoenix CLC (in front of the mural Bob painted). 1998.

Children of *Challenger* and *Columbia* together after the memorial service for the astronauts lost on the *Columbia*. Houston, Texas, February 2003.

American Institute of Aeronautics and Astronautics (AIAA) award presentation. From left are Roger Simpson, chairman of AIAA Foundation board of trustees Dr. June Scobee Rodgers, founding chairman, Challenger Center for Space Science Education; and David Thompson, AIAA president. May 12, 2009.

From left: Lucy Hawking, Richard Garriott, and June Scobee Rodgers
at the House of Commons. London, England, 2008.

June with Richard Garriott during Zero-G experience in a plane
above Austin, Texas, May 2008.

"My favorite me in Zero-G." June flying in the Zero-G aircraft, 2008.

Red and Rover comic by Brian Basset shows three sets of stars representing
Apollo, *Challenger*, and *Columbia* astronauts lost in space accidents.
Used with the artist's kind permission. This inspired the Cosmic
Rendezvous Gala to honor memories of three different crews.

STS-118 downlink from Space with Barbara Morgan.
June and students take turns asking Barbara Morgan questions that
she demonstrates while aboard the Space Station. Alexandria, Virginia, 2007.

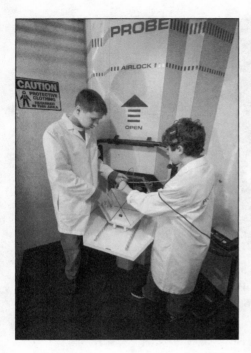

Challenger Learning
Center Space Station
Simulator: Deploying the
Probe

Challenger Learning Center Space Station Simulator:
Solving Life Support Problems

Challenger Learning
Center Space Station
Simulator: Technology in
ISO Chamber

Challenger Learning Center Space Station Simulator:
Serious Navigation Decisions

June and Sharon McKee at ten-year Challenger Center Gala. Sharon and her husband, Ellsworth, received the *Challenger 7* Award. 1996.

June with Robert Schuller at the Crystal Cathedral to present *Silver Linings* during *The Hour of Power* program. 1996.

June and Don at Guideposts Conference. Houston, Texas, 2007.

Left to right: June Scobee Rodgers; Father George Coyne, president, Vatican Observatory Foundation; Cardinal Giovanni Lajolo; The Holy Father Pope Benedict XVI; and Father Jose Funes, director, Vatican Observatory. The Vatican, August 2009.

June and George H. W. Bush. Houston, Texas, May 2010.

Kathie Scobee Fulgham and
"Muse Mother" June. Chattanooga,
Tennessee, Christmas 2008.

Colonel Rich Scobee with his wife,
Alene, in front of F-16. Ft. Carswell,
Ft. Worth, Texas, 2008.

June's 60th birthday at Disneyworld with Don and grandchildren.
The younger grandchildren, impatient to see Mickey Mouse, didn't
appreciate having to sit for a picture! 2002.

APPENDIX

THE SPACE SHUTTLE *CHALLENGER* TRAGEDY ADDRESS

Delivered January 28, 1986
President Ronald Reagan

Ladies and Gentlemen, I'd planned to speak to you tonight to report on the state of the Union, but the events of earlier today have led me to change those plans. Today is a day for mourning and remembering. Nancy and I are pained to the core by the tragedy of the shuttle *Challenger*. We know we share this pain with all of the people of our country. This is truly a national loss.

Nineteen years ago, almost to the day, we lost three astronauts in a terrible accident on the ground. But, we've never lost an astronaut in flight; we've never had a tragedy like this. And perhaps we've forgotten the courage it took for the crew of the shuttle; but they, the *Challenger* Seven, were aware of the dangers, but overcame them and did their jobs brilliantly. We mourn seven heroes: Michael Smith, Dick Scobee, Judith Resnik, Ronald McNair, Ellison Onizuka, Gregory Jarvis, and Christa McAuliffe. We mourn their loss as a nation together.

For the families of the seven, we cannot bear, as you do, the full impact of this tragedy. But we feel the loss, and we're thinking about you so very much. Your loved ones were daring and brave, and they had that special grace, that special spirit that says, 'Give me a challenge and I'll meet it with joy.' They had a hunger to explore the universe and discover its truths. They wished to serve, and they did. They served all of us.

We've grown used to wonders in this century. It's hard to dazzle us. But for twenty-five years the United States space program has been doing just that. We've grown used to the idea of space, and perhaps we forget that we've only just begun. We're still pioneers. They, the members of the *Challenger* crew, were pioneers.

And I want to say something to the schoolchildren of America who were watching the live coverage of the shuttle's takeoff. I know it is hard to understand, but sometimes painful things like this happen. It's all part of the process of exploration and discovery. It's all part of taking a chance and expanding man's horizons. The future doesn't belong to the fainthearted; it belongs to the brave. The *Challenger* crew was pulling us into the future, and we'll continue to follow them.

I've always had great faith in and respect for our space program, and what happened today does nothing to diminish it. We don't hide our space program. We don't keep secrets and cover things up. We do it all up front and in public. That's the way freedom is, and we wouldn't change it for a minute. We'll continue our quest in space. There will be more shuttle flights and more shuttle crews and, yes, more volunteers, more civilians, more teachers in space. Nothing ends here; our hopes and our journeys continue. I want to add that I wish I could talk to every man and woman who works for NASA or who worked on this mission and tell them: "Your dedication and professionalism have moved and impressed us for decades. And we know of your anguish. We share it."

There's a coincidence today. On this day 390 years ago, the great explorer Sir Francis Drake died aboard ship off the coast of Panama. In his lifetime the great frontiers were the oceans, and a historian later said, "He lived by the sea, died on it, and was buried in it." Well, today we can say of the *Challenger* crew: Their dedication was, like Drake's, complete.

The crew of the space shuttle *Challenger* honored us by the manner in which they lived their lives. We will never forget them, nor the last time we saw them, this morning, as they prepared for the journey and waved good-bye and "slipped the surly bonds of earth" to "touch the face of God."

Thank you.

NATIONAL CHALLENGER CENTER DAY, 1987

By the President of the United States of America, A Proclamation
Ronald Reagan

Will America continue to lead the world in space exploration as we move into the twenty-first century?

The *Challenger* crew, lost one year ago on the twenty-fifth space shuttle mission, dedicated themselves to America's leadership in space exploration. That leadership depends not only on our courage and determination, but also on the knowledge, capability, and inspiration of our students who will be the researchers and the astronauts of the twenty-first century

A goal of the Space Shuttle *Challenger* mission was to bring the study of space science directly and dramatically into the nation's classrooms.

In recognition of the critical need to provide America's students with access to outstanding space science education and to motivate study and excellence in science, the families of the *Challenger* crew established a Challenger Center for Space Science Education. This Center will honor the memory of the *Challenger* crew with an ongoing monument to their achievements, to their courage, and to their dedication to future generations of space explorers.

In commemoration of the brave members of the *Challenger* crew the Congress, by Senate Joint Resolution 24, has designated January 28, 1987, as "National Challenger Center Day" and authorized and requested the President to issue a proclamation in observance of this event.

Now, therefore, I, Ronald Reagan, President of the United States of America, do hereby proclaim January 28, 1987, as National Challenger Center Day and I call on the people of the United States to observe this day by remembering the *Challenger* astronauts who died while serving their country and by reflecting upon the important role the Challenger Center will play in honoring their accomplishments and in furthering their goal of strengthening space and science education.

In witness thereof, I have hereunto set my hand this twenty-eighth day of January, in the year of our Lord nineteen hundred and eighty-seven, and of the Independence of the United States of America the two hundred and eleventh.

CHALLENGER CENTER NATIONAL AWARDS DINNER (1989)

George Bush, who had been elected president, continued to pledge his support. His remarks summarized my determination to continue in a new direction of reaching for the stars.

Remarks by George H. W. Bush

The mission of the Challenger Center is very close to our hearts and close to our vision of an America that gives its children the best possible education for their challenging future. The mission of Challenger Center is to spark in our young people an interest and a joy in science, a spark that can change their lives and help make American enterprise the envy of the world. We applaud your private efforts.

We are a nation of trailblazers, risk-takers, and dream-makers. The Challenger Center believes that the future is bright, if we Americans are dedicated to making it so. That dedication is what we see in the remarkable programs of the Challenger Center.

In January 1986, Barbara and I grieved for the *Challenger* families, as did all of America. The courageous *Challenger* astronauts were very much on my mind earlier this year when we gathered on the twentieth anniversary of the moon landing to chart a course for the new century: first, space station *Freedom*—our critical next step in all our space endeavors—then back to the moon, back to the future, and this time back to stay, and then a journey to another planet. In the next century, men and women will travel to Mars.

I said then that our fallen astronauts have taken their place in the heavens so that America can take its place in the stars. Today, we are moved by the continuing faith and commitment that have moved America's space program forward and brought the Challenger Center into being.

We salute you tonight for all that you have done for our common future, and we pledge you our continuing support as you move forward with the vision and the determination to carry out the *Challenger* mission.

LETTER OF SUPPORT

Shortly after Christmas 1988, I received this letter from Senator Jake Garn. He had joined Vice President George Bush and Senator John Glenn to be with the families the day of the Challenger accident. I opened the envelope to find a delightful note. I treasure it not only for its contents, but also for the humility it has taught me.

Dear June,

All of us, I suppose, at one point or another in our lives, wonder what we might be able to accomplish that would truly serve as a legacy—what single contribution we could make that helps others for which we'd like to be remembered. Most people probably never fully find the answer to that question or don't have the opportunity to see the results of their efforts come to fruition. You are one who has been blessed with both the knowledge of your purpose and the opportunity to see it become a reality

As with all great accomplishments, what you have done with the Challenger Center did not just happen. It took an incredible amount of courage, commitment, devotion, dedication, talent, ability, and perseverance on the part of a great many people. And you, as the leader of the *Challenger* families' efforts to more than memorialize your lost loved ones but to institutionalize their mission and their collective dreams, must he given so much of the credit for their success.

I will never forget the feelings you inspired in me as I stood in the crew quarters with Vice President Bush and John Glenn and listened to you as you stood and spoke for all of the families. There you were, only hours after the accident, and in the midst of your own personal tragedy, speaking with such courage and conviction about the need for the space program to go on and for the mission of that flight to continue.

Had you said or done nothing else beyond that moment, I would have counted you among the most unforgettable, inspirational people I have ever met. But that selflessness and courage and the vision you shared carried you beyond that moment, beyond the memories and into the future. It has enabled you to bring to maturity the Challenger Center program, which offers opportunities for young people that will completely change their lives and will change the direction of the nation those young people will one day lead. Because of that, you and the other family members have literally helped shape the destiny of this great nation, and you deserve the highest commendation and the greatest possible applause for what you have accomplished.

I know there is more to be done, and I know you will continue to be actively involved in it. But you are equally deserving of some time for yourself and your family. I am so happy that you have found a companion to love and share the rest of your life with. As you know, I speak from very direct and personal experience, and I know how much more rich and full and meaningful your public life can be when you once again have someone in your personal life to share it with.

Please accept my most sincere congratulations and appreciation for everything you have done. You will always be an inspiration to me.

Sincerely,
Jake Garn
United States Senator

CHALLENGER CENTER PRESIDENTIAL AWARD

Since the day he first joined us at the crew quarters at Kennedy Space Center in Florida, President Bush has kept his promise to answer our call when we needed him. During the ceremony, we presented the Challenger Center Presidential Award to the Bushes.

Presented to President and Mrs. George H. W. Bush

Barbara Bush: Thank you very much, June and all the members of the extended *Challenger* family, for this special honor; and thank each of you for the great welcome that you have given us tonight.

I can't help but feel that this is backwards; George and I should be sitting in the audience and congratulating you. Will Rogers once said, "We can't all be heroes because somebody has to sit on the curb and clap as they go by." Well, that's what we're doing tonight. We're clapping long and hard for all of you, for everything you have accomplished in ten short years.

Like most Americans, I remember exactly where I was on that fateful day in January 1986: a hotel room in California, sitting and watching television in stunned silence. While you, the members of the *Challenger* family, struggled to deal with your own personal grief, the entire nation was gripped with a great sense of sadness and tragedy.

A Greek philosopher once said, "The test of courage is not to die, but to live." You met that great test with incredible strength and determination. Then you chose to do much more, to bring great meaning to your life and to the lives of those lost aboard the *Challenger* by turning grief into a positive force. You helped us heal the wounds much more than we were able to help you.

June, you truly are one of my heroes. As the founding chairman of the Challenger Center, you worked tirelessly to make this dream possible. At a very difficult time in the history of our country, you were the backbone for us all, and for that, we'll always be grateful.

A few years ago, when George was president, I visited one of the Challenger Learning Centers: the Howard B. Owens Science Center in Greenbelt, Maryland. What a wonderful place! It was full of knowledge, enthusiasm, and fun. Through the Challenger Learning Centers, now opening all over the country, you have given the children of our country a great gift: the gift of hope and the gift of learning.

Much has happened in the space program and in the world in the last ten years—some of it good, some of it bad. But just think: As we stand here tonight, an American astronaut is orbiting the Earth with a group of Russian cosmonauts, working together in friendship and peace.

You definitely can take some of the credit for that huge achievement . . . . It was your courage and determination that helped us all keep the dream of exploring outer space alive.

Thank you so much for making us a part of this very special evening.

George H. W. Bush: Nine years ago, an inspiring idea blossomed in the midst of great pain and sorrow. The idea was in a special way a powerful expression of a uniquely American trait: the ability to forge triumph from adversity.

The creation of the Challenger Center for Space Science Education and the founding of Challenger Learning Centers across the country were remarkable demonstrations of strength and goodwill and an abundance of faith. Just as remarkable has been the good work that has come out of these centers of hope and learning and inspiration. I think it's wonderful seeing these kids here tonight.

So tonight we celebrate the continuation of a dream and the power of an idea. Through their work, the people of the Challenger Learning Centers have taught thousands of youngsters the values which have, for more than two centuries, made our country strong and vibrant. Those who have been touched by a Challenger Learning Center have learned about teamwork, about challenge and response, about dealing with adversity and learning to solve problems, and perhaps most importantly, about the fundamental lessons of human interaction.

It has been said before, but it certainly bears repeating—especially tonight: the dream is alive. And you have made it grow.

The hopes and dreams of the *Challenger* crew were as varied as their backgrounds. They came from many places and more experiences, but they came together as a team, united by the power of an idea so simple and yet, so powerful: to teach.

Along the way, they hoped to inspire and make real in the hearts of our young people the certainty that dreams do come true, that adventure and exploration and discovery await those who take the time to prepare and have the will to make it happen.

Dick Scobee, Mike Smith, Greg Jarvis, Judy Resnik, Ron McNair, Ellison Onizuka, and the remarkable schoolteacher, Christa McAuliffe, continue to live on in our hearts. We honor their memory tonight; but moreover, the Challenger Learning Centers honor their memory—what they stood for—every day they open their doors.

You've come a long way in those ten years, but challenges lie ahead. As I see it, it boils down to two great challenges. The first is to continue to help the Challenger Centers grow. From its first learning center in Houston in 1988, the learning centers have spread to twenty-five cities. More than 600,000 students, teachers, and parents have been a part of Challenger Center programs; and thousands more have benefited from the teacher training programs. We must continue to nurture these efforts.

The second task is perhaps a more daunting one—but no less important. It is to see that the spirit of exploration and discovery continues to endure in this country. It's to make sure that American space exploration continues to push back the frontier on into the next century

Of course, it won't be easy. The space program is confronted with its own problems these day . . . adversity, technical problems, the drive to balance the federal budget. . . . As president, I never wavered in my support for the space station *Freedom*. We kept the project fully funded and on schedule, and though there have been setbacks, I hope we won't abandon it now. We also defined the new frontier in space exploration when we committed this nation to going to Mars. We raised the bar a notch, and we did so with the pride and confidence that we could reach that goal.

Never before in our nation's history has America surveyed a frontier and then retreated. We have always been a people among whose basic urges is to seek out the distant horizon, to surpass it, to go to the next, and the next after that. We are the seekers, the explorers. We are the leaders. It is the stuff of our history—the stuff of our past. I believe it should also have a place in our future.

The *Challenger* Seven lived in vibrant pursuit of a dream. As long as we continue to pursue that dream, as long as we help it to touch the lives of our young people, as long as we help to ensure that America continues to rise to the challenge of the new frontier, then it can be said that we never truly lost those seven brave souls. They will continue to live on with us, and with the hopes and dreams of the nation.

"MY MOTHER—
MY MUSE"

Kathie wrote this poem for me on Mother's Day, and her words still encourage me when I feel inadequate or insignificant. I've learned that mothering our babies and children even as they grow into their adult lives must certainly be the most creative, most challenging, and also most rewarding gift God provides.

Like dragons, mere mortals must slay
The darkness that keeps inspiration away.
Come!
Call the muses.
That lofty lot,
Who summon the gifts from the depths of the heart.

To life! Empty canvas
To mirth! Little words
Make melodies where sour notes can be heard.

But the muses are quiet.
They cannot be called.
Muses, like breezes, cannot be captured at all.

So God made Muse Mothers to love and inspire
To whisper muse words and
Muse music inside us.
To teach us the magic
In the dance of our lives
In the faith of the seasons
In the rhyme of God's sighs

So look not for your muse flying 'round in the air.
She beats in your heart.

She thinks in your head.
She coaxes the light when you founder in dread

Inspire me, Muse Mother.
Let us paint pretty thoughts.
Let us dance to the laughter.
Let us sing of love knots.

Years from now when you are musing with angels above,
I will still call you Muse Mother, my mama, my love.

Kathie Scobee Fulgham

LETTER TO THE CHILDREN OF THE *COLUMBIA* CREW

After the memorial service for the Columbia astronauts, Kathie and I and other Challenger *families met with the* Columbia *families and friends at a NASA facility at Johnson Space Center. One by one, Kathie gave a letter to each older child and parent present. Speaking from seventeen years of missing her father, Kathie wrote,*

Feb. 1, 2003
To the Children of the *Columbia* crew,

We, the *Challenger* children and all the children of public disasters, are hearing your hearts break, holding your hands, and hugging you from afar. You are not alone. Our nation mourns with you. But yours is also a personal loss that is separate from this national tragedy.

We hope this letter will bring you some comfort now or in the future, when you are strong enough and old enough to read it. We want to prepare you for what's to come and help you on grief's journey. We want you to know that it will be bad—very bad—for a little while, but it will get better. . . .

Why does the TV show the space shuttle streaking across the sky over and over again? What happened? Where is my Mom or Dad? Yours is a small voice in a crashing storm of questions. And no answers will bring you comfort.

Seventeen years ago, before some of you were even born, I watched my father and his crew die in a horrible accident. Our loved ones were astronauts on board the space shuttle *Challenger*, which blew up a few minutes after take off. It all happened on live television. It should have been a moment of private grief, but instead it turned into a very public torture. We couldn't turn on the television for weeks afterward, because we were afraid we would see the gruesome spectacle of the *Challenger* coming apart a mile up in the sky.

My father died a hundred times a day on televisions all across the country. And since it happened so publicly, everyone in the country felt like it happened to them, too. And it did. The *Challenger* explosion was a national tragedy. Everyone saw it, everyone hurt, everyone grieved, and everyone wanted to help. But that did not make it any easier for me. They wanted to say good-bye to American heroes. I just wanted to say good-bye to my daddy.

You've discovered by now that you won't be able to escape the barrage of news and the countless angles of investigation, speculation, and exasperation. The news coverage will ebb and flow, but will blindside you in the weeks, months, and years to follow when you least expect it. You will be watching television and then, suddenly, there will be that image of the shuttle—*your* shuttle—making its tragic path across the sky. For other people watching, this will all be something called "history." To you, it's your life.

Just know that the public's perception of this catastrophe isn't the same as yours. They can't know how painful it is to watch your Mom or Dad die several times each day. They can't know the horror you feel when they talk about finding your loved one's remains. If they knew how much pain it caused, they would stop.

You may have strange dreams or nightmares about your Mom or Dad being alive somehow, stranded or lost in some remote location after parachuting out of the shuttle before it flew to pieces. They may call to you in your dream to come find them. You will wake up with such hope and determination, only to have the clouds of reality gather and rain fresh tears of exasperation and sadness on your face. These dreams are your subconscious self trying to make sense out of what your conscious self already knows.

You may feel sick when you think about his or her broken body. You will be afraid to ask what happened because the answers might be worse than what you imagined. You'll torture yourself wondering if they felt pain, if they suffered, if they knew what was happening. They didn't. In the same way your brain doesn't register pain immediately when you break your arm, your Mom or Dad didn't know pain in their last moments of life on this Earth.

You will think about the last things you said to each other. You might worry that you didn't say enough or say the right things. Rest easy. Their last thoughts were of you—the *all* of who you are—not the Feb. 1, 2003, you.

And they were happy thoughts; all in a jumble of emotions so deep they are everlasting.

Everyone you know will cry fresh tears when they see you. People will try to feed you even though you know it all tastes like cardboard. They want to know what you think—what you feel—what you need. But you really don't know. You may not know for a very long time. And it will be an even longer amount of time before you can imagine your life without your Mom or Dad.

Some people, working through their own grief, will want to talk to you about the catastrophe, the aftermath, the debris recovery, or the actions that will be taken by NASA. Others will whisper as you walk by, "Her dad was killed in the space shuttle disaster." This new identity might be difficult for you. Sometimes you will want to say to the whisperers, "Yes! That was my Dad. We are so proud of him. I miss him like crazy!" But sometimes you will want to fade into the background, wanting to anonymously grieve in your own way, in your own time, without an audience.

When those who loved your Mom or Dad talk with you, cry with you, or even scream with frustration and unfairness of it, you don't have to make sense of it all. Grief is a weird and winding path with no real destination and lots of switchbacks. Look on grief as a journey—full of rest stops, enlightening sights, and potholes of differing depths of rage, sadness, and despair. Just realize that you won't be staying forever at one stop. You will eventually move on to the next. And the path will become smoother, but it may never come to an end.

Ask the people who love you and who knew and loved your Mom or Dad to help you remember the way they lived—not the way they died. You need stories about your Mom or Dad from their friends, coworkers, teachers, and your extended family. These stories will keep your Mom or Dad alive and real in your heart and mind for the rest of your life. Listen carefully to the stories. Tell them. Write them. Record them. Post them online. The stories will help you remember. The stories will help you make decisions about your life—help you become the person you were meant to be.

Please know that we are with you—holding you in our hearts, in our minds, and in our prayers.

With love,
Kathie and the children of the *Challenger* crew

1986–2006:
THE TWENTIETH
ANNIVERSARY

A Twentieth Anniversary Dinner, Houston, Texas

In 2006, we marked twenty years since the loss of our loved ones, but more important, we honored their scientific and educational legacy that lives on each day when children fly a space mission at a Challenger Learning Center. It was a spectacular evening of food, music, and celebration for two decades of Challenger Center success.

When I stepped to the podium to make a presentation, I recognized the faces of loving friends who believed in us and in our mission.

Hello and welcome. I am delighted to see all of you here tonight, as we mark the journey we have traveled together to build and grow the Challenger Center for Space Science Education.

For me and many others in this room, this is both a joyous and solemn event. This year we mark the twentieth anniversary of the Space Shuttle *Challenger* accident, when we lost seven wonderful crew members, including my former husband, Dick Scobee.

I will always feel sadness when I remember that day, but there also is reason to celebrate tonight. Together, as a team, everyone in this room and many others have contributed to the incredible success of the Challenger Center for Space Science Education. Your remarkable efforts have made this organization a living, breathing tribute to the beloved *Challenger* crew.

Tonight, I can't help but remember a conversation I had with my husband, Dick, after he returned from his first mission to space in April 1984, twenty-one months before the *Challenger* accident. He talked with wide eyes and a big grin about the wonders of the universe and everything he had experienced so many miles from home. Every time I enter a Challenger Learning Center, I see that same sense of wonderment on the faces of the young students intensely involved in a simulated space mission.

During that same conversation with Dick, I asked if it bothered him that earlier in the day, President Reagan had publicly thanked the STS 41-C crew by name but had accidentally forgotten to mention Dick. He said, "No, it doesn't bother me, because it's not about me, it's about the mission. What is important is getting the job done and having a successful mission, and we did that."

One year, nine months later, the *Challenger* 51-L crew, including the country's first teacher in space, was equally focused on its mission. And today, by staying focused on *our* mission—which is to continue to grow and enhance the Challenger Center for Space Science Education—we are honoring the crew's quest for knowledge and love of learning.

The students who walk through the doors of a Challenger Learning Center are this country's future. They are tomorrow's astronauts, engineers, and teachers. These centers help students learn that science and math can be interesting and fun, and they teach students the value of teamwork.

There is a lot to be thankful for as we look back at the success of this organization over the past twenty years. So many of you have played an integral role in the organization's success that it would be impossible for me to stand here tonight and thank only a few. All of you—Challenger families, government officials, business leaders, the NASA family, others in the science community, educators, local communities, individual volunteers—are making a difference in the lives of young people.

On Jan. 28, 1986, people around the world were saddened by our national—and, for many of us, personal—tragedy. But from that tragedy has risen hope. Thank you, from the bottom of my heart, for carrying on the *Challenger* crew's educational mission and for helping create a bright future for our youth and this country.

Statement from the *Challenger* Families

For those who were unable to attend the twentieth anniversary dinner, we wrote an op-ed piece for the Houston Chronicle *to thank the people of Space City Houston and the NASA family at Johnson Space Center.*

January 28, 1986, is a day in history that stands out as one of excitement, tragedy, and remembrance. On that day, *Challenger* 51-L, the "Teacher in Space" mission, launched into space carrying teacher Christa McAuliffe, Commander Dick Scobee, Pilot Mike Smith, and astronauts Judy Resnik, Ellison Onizuka, Greg Jarvis, and Ron McNair. Thousands of school children and citizens were watching with anticipation the launch of this mission that had captured the excitement and awe of the nation only to see a major space tragedy before their eyes. It was truly a sad day in history, but with determination and vision, our families turned this tragedy into a monumental educational opportunity for children and adults alike.

In April of that same year, we met in the living room of June Scobee, widow of Commander Dick Scobee, to discuss a memorial for our loved ones. Choosing not to have a monument in stone but rather something that would continue the education mission of 51-L, we chose to create an educational organization to inspire young people across the nation. Gathering educators, scientists, astronauts, and leaders in business and industry from across the United States, we were able to create Challenger Center for Space Science Education, which became a network of Learning Centers that uses space as a motivator to inspire students to succeed in mathematics, science, technology, and engineering while giving teachers an opportunity to change the way they teach.

These Learning Centers provide simulations in which students climb aboard a space station and work in teams to solve problems as astronauts and mission controllers in scenarios that take them through a comet, to the surface of the Moon, to a rotating platform to observe Earth, and to Mars. New scenarios are being developed to take education into new realms of excitement, creating tomorrow's prepared work force. Recognized by the United States Department of Education for being a top motivator in mathematics, science, and technology, Challenger Center embraces the National Science Education Standards in each of its missions so that teachers are receiving hands-on, minds-on delivery of concepts that must be taught.

As we begin our historic twentieth year, our Learning Centers will honor the crew and commemorate their lives and legacies throughout the year in a myriad of events. Through the growing network of fifty-one Challenger Learning Centers in the United States, Canada, and the United Kingdom, the mission of the crew truly lives on. And as new centers open, we will celebrate those communities who, like our families, come together from diverse backgrounds and experiences to create opportunities to enrich and expand the education of students in a unique and fun approach. Together our Learning Centers touch lives and create opportunities for students, our future leaders, who are the true continuation of the mission of the *Challenger, 51-L.*

As then Vice President George Bush stated in a letter to June Scobee, "The *Challenger* Seven lived in vibrant pursuit of a dream. As long as we continue to pursue that dream, as long as we help to touch the lives of young people, as long as we help to ensure that America continues to rise to the challenge of the new frontier, then it can be said that we never truly lost those seven brave souls. They will continue to live on with us, and with the hopes and dreams of the nation."

Memorial Speech at the Kennedy Space Center

Invited to speak by the Memorial Foundation at the Kennedy Space Center on January 28, 2006, I chose as the title of my speech "To Celebrate Their Legacy." Standing at a podium near the memorial in the warm Florida sunshine, I looked out to a wave of smiling faces, many who had worked alongside Dick Scobee for years.

Thank you to the Astronaut Memorial Foundation for the honor to speak—and to you who have joined me today in remembrance of our loved ones, your friends.

We all lived in Houston, in Clear Lake City, more than twenty years ago. The Scobee children worked at odd jobs baby-sitting, tossing newspapers. The Smith and Onizuka school-age children were neighbors, involved in water sports, soccer, and even a lawn mowing business. The McNair children were toddlers. On January 28, 1986, we were all together with the McAuliffe children here at KSC, waiting for the launch of *Challenger* 51-L "Teacher in Space" mission. Our lives were shattered that day, but in the years that followed, the families persevered with tremendous success. And I believe those parents lost aboard *Challenger* would be ever so proud of their children.

All the children went on to college; they represent careers in science, public relations, law, education, engineering, business, and the military. Eight have married, with thirteen grandchildren among them . . . .

That's a legacy of tremendous pride.

Sadly, the *Columbia* families have loved ones' names engraved along with those from *Apollo* and *Challenger* . . . on the Memorial Mirror here at this NASA facility. Their children are represented here today by Laura and Matthew, and we share in the pride of watching these children grow up with delight.

After the *Columbia* space shuttle accident, our daughter Kathie . . . wrote a letter from the *Challenger* children to the children of *Columbia* . . . .

Kathie said, "We could hear their hearts breaking."

It helps when a nation mourns with you, but it also hurts when a gaping wound of grief becomes so public as we were reminded again with the loss of the *Columbia* crew.

Kathie wrote a letter to their children that reminded them, "the Challenger children are hearing your hearts break and hugging you from afar. You are not alone. Our nation mourns with you. But yours is also a personal loss that is separate from this national tragedy. We hope this letter will bring you some comfort now or in the future. We want to help you on grief's journey."

Kathie is a communications manager for the Tennessee Aquarium in Chattanooga and is here today with her husband Scott and children Emily, Courtney, and Jilly.

Dick Scobee treasured time with his children . . . and would have loved all our grandchildren.

You may ask, "Why space exploration?"

Technology advances, even philosophical advances . . . just to be able to rise from the Earth, to dream of the journey to other planets and through an infinity of galaxies—all this enlarges the human horizon.

Personally, I believe God invites us to explore the depth and breadth of His Universe. He gives us curiosity and challenges us to discover . . . to learn about our place among the stars.

Their legacy . . . though their lives were taken, their spirit was not lost— for their legacy lives on—their memory as individuals and as the mighty Challenger crew . . . .

At last count, eighty-five schools, museums, planetariums, airports, auditoriums, bike paths, and streets are named for *Challenger*, McAuliffe, Smith, Jarvis, Onizuka, Resnik, McNair, and Scobee.

And the Challenger Center for Space Science Education continues to grow with a recent launch of the 53rd Challenger Learning Center at Wellington, Kansas. Others reach across our nation from California to New

York, Florida to Washington, up to Canada, across the Atlantic to England and even around the world to South Korea.

So, you see, their dreams did not die. Their mission to inspire, to explore and to learn lives on with every new space flight, and each time a child looks to the stars and wonders, "What if. . ."

Twenty years ago just after the accident the families were inspired to build Challenger Center, a living memorial to their lives—to their Challenger mission. The world knew how they died; we wanted the world to know how they lived. It wasn't easy. There were naysayers, but a few people believed in us.

Ten years ago today, my family celebrated the memory of our loved ones aboard *Challenger* in Phoenix, Arizona, at the Super Bowl. We were there to watch our son fly an F-16 in a missing man formation over the stadium at Super Bowl XXX. . . .

The flyover, led by Capt. Rich Scobee, son of the *Challenger* Commander, Dick Scobee, was planned to commemorate the tenth anniversary of the *Challenger* tragedy.

And now, Dr. Feldman, executive director for the Memorial Foundation asked me to introduce my son to share with you some fond memories of his Dad and their joy of life in the military and airplanes.

Our son, Rich Scobee is here today with his children, Dexter and Cristi. Today he is a Colonel in the Air Force and Commander of the 301st Fighter Squadron at Luke AFB. He is a graduate of the Air Force Academy, distinguished graduate from pilot training, graduate of the Air Command and Staff College and the Air War College. He has flown 45 combat missions over Northern Iraq, and has served as a Flight Commander, Combat Instructor and Evaluator pilot, with more than 300 hours combat time.

It is my honor and great pleasure to introduce your next speaker, my son, Colonel Richard Scobee (Scobs). . . .

"TEACHER ON SHUTTLE TO CALL LOCAL WOMAN"

Karen Nazor Hill
Chattanooga Times/Free Press
August 15, 2007

When teacher-turned-astronaut Barbara Morgan calls this morning from the space shuttle *Endeavor* to talk to students at the Challenger Center outside Washington, DC, Chattanoogan June Scobee Rodgers will answer the call.

"I've been preparing my message," Rodgers said. "It will go something like 'Congratulations, *Endeavor* crew. Barbara, we've been standing by at the Challenger Center waiting for your call for 21 years.'"

Rodgers said she has been dreaming of the day a teacher finally would make it into space.

On Jan. 28, 1986, the shuttle *Challenger* exploded, killing Dr. Rodgers's husband, shuttle commander Dick Scobee, and six other astronauts, including teacher Christa McAuliffe. Since then, Dr. Rodgers has been an outspoken advocate for sending a teacher into space.

Ms. Morgan was a back-up astronaut for Ms. McAuliffe and trained with the *Challenger* crew.

Dr. Rodgers said she is "over-the-moon happy" about Ms. Morgan's call from space.

"It's a culmination of all the rewards of the Challenger Learning Centers that were founded in honor of the education mission of the *Challenger*," she said. "It has caught on nationally and around the world. It has come full circle—millions of kids are flying (in simulated flights in Challenger Centers)."

Mrs. Morgan will field questions from 20 children in a 20-minute live interview at the Challenger Center for Space Science Education in Alexandria, Va.

Twenty-one fifth-graders at Battle Academy in Chattanooga, as well as students in each of the 51 Challenger Learning Centers, took part in an interactive Webcast Wednesday with former astronauts Dr. Joseph Allen, William Readdy and Dr. Roger Crouch, said Kaitlyn Vann, Challenger Center flight director.

Chattanooga's Challenger Center director Tom Patty said Battle Academy is a certified NASA Explorer School and was selected via an application process.

Student Shania Douglas, 10, said it was an honor to participate in the event.

"It's so cool to get to submit a question to astronauts and hear what they have to say," Shania said. "I want them to do good work up there that will help us down here on Earth."

Dr. Rodgers attended *Endeavor*'s launch last week at Cape Canaveral, Fla. It was her first since witnessing the *Challenger*'s explosion in 1986.

"Many of my friends have flown, and I've watched the launch on TV and said prayers as it was launching," she said. "But I couldn't make myself go back."

She said Barbara Morgan is the reason she changed her mind.

"I've been waiting," she said. "There was no way I would have missed this, whether it was to get there hitchhiking or crawling—nothing would have stopped me."

Dr. Rodgers said she held onto her husband, retired Army Lt. Gen. Don Rodgers, "for dear life" during the launch.

"I was on the edge of existence as the shuttle rocket engines ignited and puffs of steam came out and it lifted off the launch pad. It was a steady stream of bright orange, and it kept going higher and higher. Then it rotated and went off like a star in the sky," she said.

"Once in orbit, I broke down in tears of joy, delight and relief, and then I just buried myself in Don's shoulder and cried," she said. "I've just been so involved since the *Challenger*, and to experience this as well as recalling Dick Scobee and his crew on the *Challenger* just released all these emotions."

"WHAT I LEARNED FROM THE CHALLENGER CREW"

At the Challenger Center Annual Conference, Barbara Morgan ended her presentation the following words. When she concluded, the audience jumped to its feet in applause. (See www.challenger.org to watch her speech.)

Remarks to the Challenger Center
2009 Annual Conference, Paramus, New Jersey
Barbara Morgan

Thank you, thank you, thank you for all you do to give our young people and their teachers—rich, real-world science, math, engineering, and technology learning experiences that they will never forget.

Thank you for making an enormous difference for education and exploration. I hope you will always remember what a tremendous impact you make on our world—one young explorer, one young pair of flight controllers, one team, one class, at a time.

You really do launch dreams and change lives. I am proud of you. I am proud to be your colleague. Our journey, and our work at our Challenger Learning Centers, began many years ago, with seven wonderful people.

And for me, personally, through all those years, and through 5.3 million miles of spaceflight, I have carried what I learned from Dick Scobee, Mike Smith, El Onizuka, Judy Resnik, Ron McNair, Greg Jarvis, and Christa McAuliffe.

Today, we're here at this Awards Luncheon, to honor all of you—for your wonderful service, teaching, and leadership.

So I'd like to share with you some of the things I learned about teaching and leadership from our *Challenger* crewmates.

Our commander, Dick Scobee, was a deep thinker, a poetic thinker, and a loving person. Dick was so modest that even now it is still hard to praise him without thinking I'll embarrass him.

But leaders are teachers, and Dick taught me that a true leader guides more than he or she commands. A true leader seems to stand beside us rather than ahead of us. And then, when we arrive at our goal, we realize he was already there, before us.

Dick was there with us even when the training was over for the day. Dick and June Scobee welcomed us as members of their family. They made us feel that we would be members of their family forever. They welcomed us at suppertime and at Christmas time. Until the early hours of the morning, we sat on their living room carpet and talked about the wonders of flight and space. We talked about the wonder of children and how the prospect of exploration and discovery can motivate them for the rest of their lives.

Dick taught me that it is motivation that makes one capable, and that if we can motivate our children, *everything* is possible.

From our shuttle pilot, Mike Smith, I learned that a great leader leads by *trusting* the people he or she works with—the same way that a great teacher teaches by *trusting* his or her students. In a high-performance T-38 jet trainer, Mike took me on the most exciting flights I had ever had in my life. He showed me how to fly in formation. He showed me what it's like to break the sound barrier. He showed me how to do barrel rolls and lazy eights. Then he told me it was my turn.

Mike said, "Push the stick."

When I asked, "Which way?" he said, "Any way you want. Push the stick."

And so I flew the T-38, taking it through a series of barrel rolls and long sweeping arcs high above the Gulf of Mexico.

Before the flight, I had worried whether I could handle the oxygen mask and ejection seat, but Mike had more confidence in me than I had in myself. He shared with me the job he loved so well. He helped me open a new door of opportunities. (Of course, at that time, back then, Mike also had his hand on the stick at all times!)

From Judy Resnik, I learned that you don't have to try to be all things at once. Judy flew third seat on the *Challenger*. She sat right between and behind Dick and Mike. She monitored everything they did in flying the shuttle. She was always on top of the situation.

On the job, Judy was all concentration. Off the job, she was fun and she was a friend.

Judy also taught me that a great teacher and leader can command very technical information and still make that information understandable to her students. With Judy as my teacher, back then, even I could understand orbital mechanics!

From Ellison Onizuka, I learned that the best teacher teaches by learning and by allowing his students to learn alongside him. El invited Christa and me to his classes so that we could see *how* he learned so that we could learn *what* he learned.

One of the training sessions we attended with El was his "Onboard Television Malfunctions" class. We didn't want to get in his way, so we stood back in the corner of the simulator. But El had us move over and stand right next to him.

As El and his trainer went through a series of malfunction exercises, El took the time to explain each step to us. It didn't matter to El that he really didn't have the time—he made the time.

At the end of the session, El made the time to go through a couple of the exercises again, only this time he asked each of us to take his place. He had Christa and me each go through his malfunctions checklist, find the problems, and fix them.

Classes like El's were not a standard part of our training as space flight participants. But, like the other crew members and all good teachers, El knew the importance of teaching beyond the basics.

From Ron McNair, I learned faith. Strong faith. Faith in a higher being, faith in oneself, faith in others. Ron showed me how to be accepting of all others.

He showed me that a great leader leads by caring, and that one way to show you care is to learn about the people you lead.

Ron often talked to Christa and me about our homes and our families. He knew about our schools, our students, and our other interests. He also spent time with the three students whose experiments he was going to conduct onboard the shuttle. He treated their experiments as professionally as he treated every experiment on the mission.

From Greg Jarvis, I learned that no matter how many times you face disappointment, you keep going and going until you reach your goal.

Greg taught me that a great leader leads by example, and he set a great example for the whole crew.

As a payload specialist, Greg was working on fluid dynamics and satellite projects for Hughes. Because of scheduling reasons, Greg had been "bumped" from several earlier flights before flying on the *Challenger*. One time, he was bumped just one month before his crew was scheduled to launch.

Despite these disappointments, Greg joined the crew—bright, eager, and excited to work with everyone. He became a valuable team member immediately, even though the rest of us had been training together for months. He helped the crew both technically and psychologically. As soon as he appeared, it was obvious that Greg Jarvis would make the flight more successful.

From Christa McAuliffe, I learned to look for the best in all situations and people. She taught me not to worry about what is *not* important and at the same time to work hard at what *is* important.

What was important to Christa? People and their dignity. Her classroom was based on mutual respect. That "mutual respect" meant that her students had respect for themselves, too. She taught her students to do the best they could do and to be true to inner selves.

And she set a great example. Christa was true to herself and to our profession. She championed her students, their education, and their future.

Before the *Challenger* launch, Christa had a lot of things to do. She was training for space flight. She was preparing her lessons. She was working with the media. She was balancing all her responsibilities.

But Christa was also still working with her students.

Up through the last day, just before the *Challenger* launch, Christa took time out in the crew quarters at Kennedy Space Center to write recommendations for her students. She wrote recommendations for her straight-A students, like Andy, who really didn't need any recommendations. And she wrote recommendations for her "other" students—like "Dan," I'll call him—whom Christa said needed all the help he could get.

At a time when many people would think only of the impending launch, Christa was taking care of a teacher's business.

Well, those seven wonderful people all sound like you, don't they? Thank you for carrying on their dreams. Thank you all so much.

CHALLENGER LEARNING CENTER SUCCESSES: IN THEIR OWN WORDS

My experience with the Challenger Learning Center has had a profound influence in shaping my career path. I first attended The Discovery Museum CLC during a middle school class trip. I already had an interest in science, but I was blown away by the immersive power of the program. I still remember standing in breathless anticipation as we waited for the probe to send back data from the comet. The real connection, however, came after I began volunteering at the museum that summer under the guidance of my flight director (and CLC program director at the time) John Labate. Within a short amount of time, I found myself on the other side of the flight commander console, as well as teaching and developing outreach programs and school classes.

The CLC program allows students to learn because they are excited and want to be able to complete their task, rather than forcing knowledge in preparation for a test. It affords a brief glimpse for the participants into what makes science cool and exciting. In almost every class, you can see that at least one student really connects with the program. But for all the students who I directed through the CLC, I think I was the one who benefited most from this hands-on science approach, as my time at the museum provided unique on the job training for a young scientist—not only in how to teach fundamental concepts in an engaging way, but in the technical experience of exhibit and demonstration troubleshooting and construction. It served as a crash course in experimental science and intuitive thinking that continues to serve me well in my career. Working at the museum allowed me to keep a real, tangible hold on the fun of science and discovery, even during periods in my own schooling when I was otherwise turned off from science. I firmly believe that my time at The Discovery Museum and the CLC is the reason I am a scientist today.

> — *Brendan Hermalyn, former CLC student, currently enrolled in a doctoral program (Ph.D.) at Brown University with a concentration in Impact Physics/Planetary Science*

I believe the Challenger Center can inspire young students to pursue a career in medicine because it brings an element of excitement to science and technology. Science plays a vital role in the medical field. It allows us to develop new medications, treatments, and surgical procedures on a daily basis. Facilities like the Challenger Center plant a seed of curiosity into the minds of our future doctors, nurses, and researchers.

> —*Emily Brewer, FNP-BC, former Challenger Learning Center student*

I've always had a passion for space, but I never knew how my universe would be turned upside down. On May 16, 2006, I was walking into my middle school locker room when I was sucker-punched by a bully. Entirely out of breath, I was brought to my knees. Two days later, I was in the hospital— paralyzed from the waist down. Prior to the incident, I had performed commercials and jingles professionally in New York City as well as attended a performing arts summer camp. After the incident, I was unable to return to that camp. My mother had always remembered my passion for space. So, while looking locally for another camp we came across the Lower Hudson Valley Challenger Center in Airmont, New York. My mom had asked the director, John Huibregtse, if I would be able to attend. As soon as he said yes, I figured I'd rediscover my inner space geek and give it a try.

After one week of fun and space simulations, I was hooked. I was just at the age limit of being able to continue as a camper, so I decided to volunteer. Wheelchair and all, my boss, John Huibregtse, and fellow flight directors were all very welcoming and accommodating to whatever needed to be done. As well, they ignored the chair and mainly focused on me for who I was. I had found a sanctuary. Finally, on February 18, 2008, my fourteenth birthday, I was employed at the center. I will admit to this day that was one of the most exciting moments of my life and the Challenger Center has opened so many doors for me. I went from a little boy in a wheelchair to Commander Sawyer.

Even now, whenever I feel down or am having a bad day, I know that I can always look to the Challenger Center, because I also realize that when I go there, I'm not just making myself happy, but everybody that walks through the door and to see their reactions to the experience itself is priceless. It is so great to go in and work with children, the future of our society, and hopefully make an impact on their lives as much as the center made an impact on mine.

Since then, many other doors have been opened and many experiences have been afforded to me. I have been able to see a space shuttle launch in person; I have met Neil Armstrong and Buzz Aldrin, and I have become known throughout the Twitterverse as @thenasaman, even hosting a space-themed podcast called "Talking Space" with more than 10,000 subscribers. However, no matter what, the Challenger Center has been the greatest discovery of my life. In essence, Challenger Centers saved my life.

—*Sawyer Rosenstein*

Challenger
C E N T E R

To: The Challenger Learning Center Network
From: June Scobee Rodgers
Date: 3 March 2009
RE: Message from Astronaut Buzz Aldrin

I know one good answer to the questions people are asking about American education, American scientific leadership, and American students' motivation to learn. That answer is Challenger Center.

Buzz Aldrin, NASA Apollo Astronaut

INSPIRE, EXPLORE, LEARN:
REAFFIRMING AND EXPANDING *CHALLENGER'S* MISSION AFTER TWENTY-FIVE YEARS

Our Challenger Center vision is a culmination of wishes and dreams from years past, but with a perceptive eye on relevant education needs of the future. Today the leadership of our board of directors and staff has amassed a talented group of people who not only believe in the mission but who are willing to work hard to make our dreams come true. The following message reaffirms our mission. As former astronaut and senator John Glenn has said, "Inspiring. Exploring. Learning. It's our mission. That's hard to beat!"

Dan Barstow
President, Challenger Learning Center

Such a clear and compelling mission. *Inspire* people to dream and achieve. *Explore* beyond the known and familiar. *Learn* as we go and help others to learn with us.

Since the tragic loss of the *Challenger* space shuttle and its crew of seven heroes twenty-five years ago, Challenger Learning Centers have embraced this mission. To date, 4 million students have participated in our simulated space missions, and we expect to reach millions more over the next decade and beyond.

Looking forward, we will transform our methods and tools, keeping up with developments in space exploration, new understandings of teaching and learning, new technologies of communication and collaboration, and new opportunities to fire the imagination and engage learners in the real world of science, technology, engineering, and math (STEM).

Yet we will remain focused on the mission that drives our soul. *Inspire, explore, learn.*

It has more importance now than ever before in our history. The *Challenger* tragedy affected our nation deeply. Not only did we lose seven fine astronauts, including the first teacher in space, we questioned our nation's drive and ability to send humans to space to explore worlds unknown. Reinvigorating our spirit of exploration became a critical issue, requiring leadership and action in all realms of society—political, business, scientific, technical, religious, educational—from the families of grieving students to the nation's president charting a new strategy for space exploration.

Our nation did successfully return the shuttle to flight. We went on to the International Space Station, launched the Hubble telescope, and sent robots to Mars. Yet we also looked to the future and knew we had to engage and inspire the next generation. Thus was born the Challenger Center for Space Science Education and the network of nearly fifty Challenger Learning Centers.

Each Challenger Learning Center has a space flight simulator. With its mission control and spaceship, students perform simulated missions to Earth orbit, the moon, Mars, and Halley's comet. Students immerse themselves in the missions, solve problems, coordinate with teammates, navigate to their destinations, complete their goals, and return safely to Earth. The experience truly changes lives as students discover the thrill of exploration and realize that STEM fields open doors to the future. And Challenger Learning Centers have become creative centerpieces of educational reform for their communities.

Now our nation again finds itself at a crisis point—in education and in space exploration. Challenger Learning Centers must rise to the occasion to meet this crucial need to inspire, explore and learn, for the good of our children, our nation, the world and the future. Although our teachers work nobly, our schools are struggling. We are falling behind other nations in educational achievement and have fewer students pursuing STEM careers. We cannot let this happen. Whether for space exploration, tapping into alternative energy sources, or making new medical discoveries, our future depends on strengthening education. We must re-invigorate learning, and we must inspire a generation of students to dream big dreams and make them real.

Challenger Learning Centers meet that vital national need to inspire, explore and learn.

Today, *Challenger's* history is prologue to the future. More than ever, our nation, indeed the world, needs us to inspire children, and help them explore and learn. We will reach well beyond the 400,000 children we now serve annually—this number should be in the millions. We will deepen the learning experiences before and after the missions and branch out into other science domains. We will expand our teacher training and after-school programs and deploy new technologies to extend our reach beyond the simulation experience. In an increasingly global community, we will expand our presence beyond our national borders.

Let us frame this as a grand challenge:

Become the world's most inspiring program for science and technology, engineering and math education.

This may sound bold and ambitious. Good. It is.

And yet it is clearly within our reach—and defines our essential contribution to the students we serve, our nation, and the world. At the core, we inspire students through direct service—now we will inspire more students in deeper ways. Our programs and services inspire well beyond our direct contact. Our students share their enthusiasm with friends and family. Teachers share with their colleagues and administrators. And our model of engaging, hands-on, inquiry-based learning inspires other programs in STEM education. All of this puts us in a world-class position of educational leadership, poised to extend and deepen our influence.

We will build on what we have accomplished over the years—our network of Challenger Learning Centers, with their wonderfully creative and committed staff, the many partners who supported our work, the underlying model of immersive simulations, and the millions of students and their teachers who have participated in our programs.

Achieving these goals will require hard work, time, energy, resources, and long-term commitment. And we will need wisdom and flexibility to adapt our pathways as circumstances change and new opportunities emerge.

We will succeed. We will carry on the mission of the *Challenger* 51L heroes—to inspire, explore and learn—well beyond even their ambitious dreams!

RELATED WEBSITES

For more about the Challenger Center programs, including our network of Challenger Learning Centers (CLC): **www.challenger.org**

For more CLC stories: **http://www.challenger.org/about/media/profiles.cfm**

For more about the Star Challengers *book series and educational resources in support of the books:*
www.StarChallengers.com
www.challenger.org/starchallengers

For more about NASA astronauts and their space flights: **www.nasa.gov/**

For more about the Challenger 51-L *mission and crew:* **www.nasa.gov/ mission_pages/shuttle/shuttlemissions/archives/sts-51L.html**

For more about Barbara Morgan's flight on the space shuttle and educational activities in support of this historic mission:
www.challenger.org/programs/sts118act.cfm

For more about Richard Garriott, his flight to the International Space Station, and educational resources in support of the flight:
www.richardgarriott.com/
www.richardinspace.com/
www.challenger.org/programs/garriottchallenge.cfm

For more about Sally Ride's educational science programs:
www.SallyRideScience.com

To help launch young people toward discovery and achievement:
http://giving.challenger.org

For more about inspiration, hope, and faith:
www.guideposts.org/
www.guideposts.com/outreach

For more about the Vatican Observatory: **www.vaticanobservatory.org/**

INDEX